AF179985

Biodiesel Production: Processes and Technologies

Biodiesel Production: Processes and Technologies

Edited by
Damian Price

Larsen & Keller
www.larsen-keller.com

Biodiesel Production: Processes and Technologies
Edited by Damian Price
ISBN: 978-1-63549-041-1 (Hardback)

© 2017 Larsen & Keller

Larsen & Keller
Published by Larsen and Keller Education,
5 Penn Plaza,
19th Floor,
New York, NY 10001, USA

Cataloging-in-Publication Data

Biodiesel production : processes and technologies / edited by Damian Price.
p. cm.
Includes bibliographical references and index.
ISBN 978-1-63549-041-1
1. Biodiesel fuels. 2. Biomass energy--Technological innovations. I. Price, Damian.
TP359.B46 B56 2017
662.669--dc23

This book contains information obtained from authentic and highly regarded sources. All chapters are published with permission under the Creative Commons Attribution Share Alike License or equivalent. A wide variety of references are listed. Permissions and sources are indicated; for detailed attributions, please refer to the permissions page. Reasonable efforts have been made to publish reliable data and information, but the authors, editors and publisher cannot assume any responsibility for the vailidity of all materials or the consequences of their use.

Trademark Notice: All trademarks used herein are the property of their respective owners. The use of any trademark in this text does not vest in the author or publisher any trademark ownership rights in such trademarks, nor does the use of such trademarks imply any affiliation with or endorsement of this book by such owners.

The publisher's policy is to use permanent paper from mills that operate a sustainable forestry policy. Furthermore, the publisher ensures that the text paper and cover boards used have met acceptable environmental accreditation standards.

Printed and bound in China.

For more information regarding Larsen and Keller Education and its products, please visit the publisher's website www.larsen-keller.com

Table of Contents

Preface

This book elucidates the concepts and innovative models around prospective developments with respect to biodiesel production. It describes in detail the various techniques and applications of this technology. Biodiesel production refers to the process of using esterification and transesterification to produce biofuels and biodiesel. Biodiesel is a substitute for diesel and can be used in a diesel engine. The book aims to provide thorough insights into this field of biodiesel production. It unfolds the innovative aspects of this field which will be crucial for the holistic understanding of the subject matter. The topics covered in this extensive text deal with the core subjects of this area. Coherent flow of topics, student-friendly language and extensive use of examples make this textbook an invaluable source of knowledge.

A detailed account of the significant topics covered in this book is provided below:

Chapter 1- Vegetable oil or diesel fuel that contains animal fat is known as biodiesel. The procedure of making biodiesel or biofuel is termed as biodiesel production. This chapter on biodiesel production offers an insightful focus, keeping in mind the complex subject matter.

Chapter 2- Fuel can be produced by geological processes or by biological processes. Biofuel is produced by biological processes and is derived directly from plants or indirectly from agriculture or domestic/industrial waste. The major biofuels discussed in the following text are cellulosic ethanol, vegetable oil fuel, ethanol fuel, babassu oil, methanol fuel and biogas. This section will provide an integrated understanding of biofuel.

Chapter 3- One of the techniques related to biodiesel production is the mixing of alcohol fuels with biomass. It is done by using MixAlco process. Biomass heating systems helps in generating heat from biomass. The aspects elucidated in this text are of vital importance, and provide a better understanding of biodiesel production.

Chapter 4- The production of biodiesel deals with a number of methods and components. Some of these methods are titration, tallow, fatty acid methyl ester and husk. Titration is a common laboratory method which is used to regulate unknown concentration of identified components. The chapter strategically encompasses and incorporates the major components and methods of biodiesel production, providing a complete understanding.

Chapter 5- The procedure of replacing organic group R" of an ester with organic R' of an alcohol is known as transesterification. The common transesterification catalysts explained in this section are sodium hydroxide, potassium hydroxide and sodium methoxide. This text has been carefully written to provide an easy understanding of the varied facets of transesterification.

Chapter 6- The essential aspects elucidated in this section are cellulosic ethanol commercialization, energy returned on energy invested, Fischer-Speier esterification and glycerol. The method of converting cellulose-containing organic matter into cellulosic ethanol for using it as a biofuel is known as cellulosic ethanol commercialization. The chapter closely examines the aspects of biodiesel to provide the reader an extensive understanding of the subject.

I would like to make a special mention of my publisher who considered me worthy of this opportunity and also supported me throughout the process. I would also like to thank the editing team at the back-end who extended their help whenever required.

Editor

1

Introduction to Biodiesel Production

Vegetable oil or diesel fuel that contains animal fat is known as biodiesel. The procedure of making biodiesel or biofuel is termed as biodiesel production. This chapter on biodiesel production offers an insightful focus, keeping in mind the complex subject matter.

Biodiesel Production

Biodiesel production is the process of producing the biofuel, biodiesel, through the chemical reactions transesterification and esterification. This involves vegetable or animal fats and oils being reacted with short-chain alcohols (typically methanol or ethanol). The alcohols used should be of low molecular weight, ethanol being one of the most used for its low cost. However, greater conversions into biodiesel can be reached using methanol. Although the transesterification reaction can be catalyzed by either acids or bases the most common means of production is base-catalyzed transesterification. This path has lower reaction times and catalyst cost than those posed by acid catalysis. However, alkaline catalysis has the disadvantage of its high sensitivity to both water and free fatty acids present in the oils.

Process Steps

The major steps required to synthesize biodiesel are as follows:

Feedstock Pretreatment

Common feedstock used in biodiesel production include yellow grease (recycled vegetable oil), "virgin" vegetable oil, and tallow. Recycled oil is processed to remove impurities from cooking, storage, and handling, such as dirt, charred food, and water. Virgin oils are refined, but not to a food-grade level. Degumming to remove phospholipids and other plant matter is common, though refinement processes vary.

Regardless of the feedstock, water is removed as its presence during base-catalyzed transesterification causes the triglycerides to hydrolyze, giving salts of the fatty acids (soaps) instead of producing biodiesel.

Determination and Treatment of Free Fatty Acids

A sample of the cleaned feedstock oil is titrated with a standardized base solution in order to determine the concentration of free fatty acids (carboxylic acids) present in the vegetable oil sample. These acids are then either esterified into biodiesel, esterified into glycerides, or removed, typically through neutralization.

Reactions

Base-catalyzed transesterification reacts lipids (fats and oils) with alcohol (typically methanol or ethanol) to produce biodiesel and an impure coproduct, glycerol. If the feedstock oil is used or has a high acid content, acid-catalyzed esterification can be used to react fatty acids with alcohol to produce biodiesel. Other methods, such as fixed-bed reactors, supercritical reactors, and ultrasonic reactors, forgo or decrease the use of chemical catalysts.

Product Purification

Products of the reaction include not only biodiesel, but also byproducts, soap, glycerol, excess alcohol, and trace amounts of water. All of these byproducts must be removed to meet the standards, but the order of removal is process-dependent.

The density of glycerol is greater than that of biodiesel, and this property difference is exploited to separate the bulk of the glycerol coproduct. Residual methanol is typically recovered by distillation and reused. Soaps can be removed or converted into acids. Residual water is also removed from the fuel.

Reactions

Transesterification

Animal and plant fats and oils are composed of triglycerides, which are esters formed by the reactions of three free fatty acids and the trihydric alcohol, glycerol. In the transesterification process, the added alcohol (commonly, methanol or ethanol) is deprotonated with a base to make it a stronger nucleophile. As can be seen, the reaction has no other inputs than the triglyceride and the alcohol. Under normal conditions, this reaction will proceed either exceedingly slowly or not at all, so heat, as well as catalysts (acid and/or base) are used to speed the reaction. It is important to note that the acid or base are not consumed by the transesterification reaction, thus they are not reactants, but catalysts. Common catalysts for transesterification include sodium hydroxide, potassium hydroxide, and sodium methoxide.

Almost all biodiesel is produced from virgin vegetable oils using the base-catalyzed technique as it is the most economical process for treating virgin vegetable oils, requiring only low temperatures and pressures and producing over 98% conversion yield (provided the starting oil is low in moisture and free fatty acids). However, biodiesel produced from other sources or by other methods may require acid catalysis, which is much slower. Since it is the predominant method for commercial-scale production, only the base-catalyzed transesterification process will be described below.

Triglycerides (1) are reacted with an alcohol such as ethanol (2) to give ethyl esters of fatty acids (3) and glycerol (4):

1 2 3 4

R^1, R^2, R^3 : Alkyl group

The alcohol reacts with the fatty acids to form the mono-alkyl ester (biodiesel) and crude glycerol. The reaction between the biolipid (fat or oil) and the alcohol is a reversible reaction so excess alcohol must be added to ensure complete conversion.

Base-catalysed Transesterification Mechanism

The transesterification reaction is base catalyzed. Any strong base capable of deprotonating the alcohol will do (e.g. NaOH, KOH, sodium methoxide, etc.), but the sodium and potassium hydroxides are often chosen for their cost. The presence of water causes undesirable base hydrolysis, so the reaction must be kept dry.

In the transesterification mechanism, the carbonyl carbon of the starting ester ($RCOOR^1$) undergoes nucleophilic attack by the incoming alkoxide (R^2O^-) to give a tetrahedral intermediate, which either reverts to the starting material, or proceeds to the transesterified product ($RCOOR^2$). The various species exist in equilibrium, and the product distribution depends on the relative energies of the reactant and product.

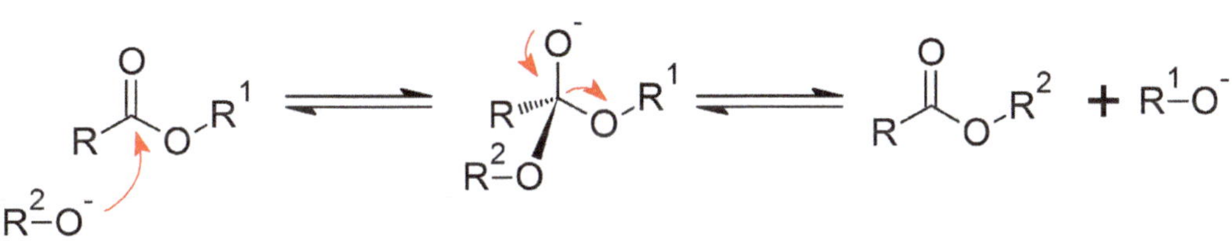

Production Methods

Supercritical Process

An alternative, catalyst-free method for transesterification uses supercritical methanol at high temperatures and pressures in a continuous process. In the supercritical state, the oil and methanol are in a single phase, and reaction occurs spontaneously and rapidly. The process can tolerate water in the feedstock, free fatty acids are converted to methyl esters instead of soap, so a wide variety of feedstocks can be used. Also the catalyst removal step is eliminated. High temperatures and pressures are required, but energy costs of production are similar or less than catalytic production routes.

Ultra- and High-shear in-line and Batch Reactors

Ultra- and High Shear in-line or batch reactors allow production of biodiesel continuously, semi- continuously, and in batch-mode. This drastically reduces production time and increases production volume.

The reaction takes place in the high-energetic shear zone of the Ultra- and High Shear mixer by reducing the droplet size of the immiscible liquids such as oil or fats and methanol. Therefore, the smaller the droplet size the larger the surface area the faster the catalyst can react.

Ultrasonic Reactor Method

In the ultrasonic reactor method, the ultrasonic waves cause the reaction mixture to produce and collapse bubbles constantly. This cavitation simultaneously provides the mixing and heating required to carry out the transesterification process. Thus, using an ultrasonic reactor for biodiesel production drastically reduces the reaction time, reaction temperatures, and energy input. Hence the process of transesterification can run inline rather than using the time consuming batch processing. Industrial scale ultrasonic devices allow for the industrial scale processing of several thousand barrels per day.

Lipase-catalyzed Method

Large amounts of research have focused recently on the use of enzymes as a catalyst for the transesterification. Researchers have found that very good yields could be obtained from crude and used oils using lipases. The use of lipases makes the reaction less sensitive to high free fatty-acid content, which is a problem with the standard biodiesel process. One problem with the lipase reaction is that methanol cannot be used because it inactivates the lipase catalyst after one batch. However, if methyl acetate is used instead of methanol, the lipase is not in-activated and can be used for several batches, making the lipase system much more cost effective.

Volatile Fatty Acids from Anaerobic Digestion of Waste Streams

Lipids have been drawing considerable attention as a substrate for biodiesel production owing to its sustainability, non-toxicity and energy efficient properties. However, due to cost reasons, attention must be focused on the non-edible sources of lipids, in particular oleaginous microorganisms. Such microbes have the ability to assimilate the carbon sources from a medium and convert the carbon into lipid storage materials. The lipids accumulated by these oleaginous cells can then be transesterified to form biodiesel.

Biodiesel

Bus run by biodiesel.

Space-filling model of methyl linoleate, or linoleic acid methyl ester, a common methyl ester produced from soybean or canola oil and methanol

Biodiesel refers to a vegetable oil - or animal fat-based diesel fuel consisting of long-chain alkyl (methyl, ethyl, or propyl) esters. Biodiesel is typically made by chemically reacting lipids (e.g., vegetable oil, soybean oil, animal fat (tallow)) with an alcohol producing fatty acid esters.

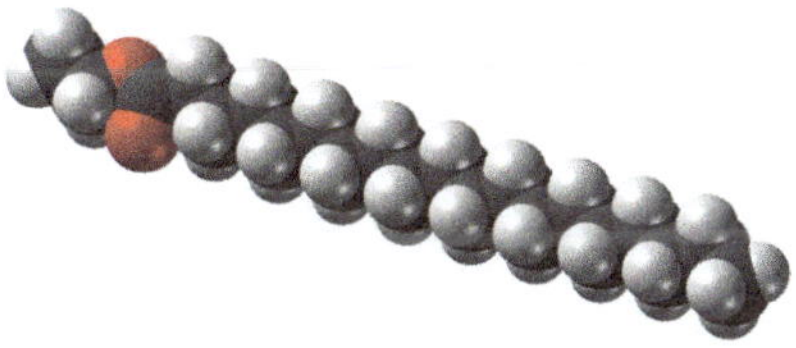

Space-filling model of ethyl stearate, or stearic acid ethyl ester, an ethyl ester produced from soybean or canola oil and ethanol

Biodiesel is meant to be used in standard diesel engines and is thus distinct from the vegetable and waste oils used to fuel *converted* diesel engines. Biodiesel can be used alone, or blended with petrodiesel in any proportions. Biodiesel blends can also be used as heating oil.

The National Biodiesel Board (USA) also has a technical definition of "biodiesel" as a mono-alkyl ester.

Blends

Biodiesel sample

Blends of biodiesel and conventional hydrocarbon-based diesel are products most commonly distributed for use in the retail diesel fuel marketplace. Much of the world uses a system known as the "B" factor to state the amount of biodiesel in any fuel mix:

- 100% biodiesel is referred to as B100
- 20% biodiesel, 80% petrodiesel is labeled B20
- 5% biodiesel, 95% petrodiesel is labeled B5
- 2% biodiesel, 98% petrodiesel is labeled B2

Blends of 20% biodiesel and lower can be used in diesel equipment with no, or only minor modifications, although certain manufacturers do not extend warranty coverage if equipment is damaged by these blends. The B6 to B20 blends are covered by the ASTM D7467 specification. Biodiesel can also be used in its pure form (B100), but may require certain engine modifications to avoid maintenance and performance problems. Blending B100 with petroleum diesel may be accomplished by:

- Mixing in tanks at manufacturing point prior to delivery to tanker truck
- Splash mixing in the tanker truck (adding specific percentages of biodiesel and petroleum diesel)
- In-line mixing, two components arrive at tanker truck simultaneously.
- Metered pump mixing, petroleum diesel and biodiesel meters are set to X total volume, transfer pump pulls from two points and mix is complete on leaving pump.

Applications

Biodiesel can be used in pure form (B100) or may be blended with petroleum diesel at any concentration in most injection pump diesel engines. New extreme high-pressure (29,000 psi) common rail engines have strict factory limits of B5 or B20, depending on manufacturer. Biodiesel has different solvent properties than petrodiesel, and will degrade natural rubber gaskets and hoses in vehicles (mostly vehicles manufactured before 1992), although these tend to wear out naturally and most likely will have already been replaced with FKM, which is nonreactive to biodiesel. Biodiesel has been known to break down deposits of residue in the fuel lines where petrodiesel has been used. As a result, fuel filters may become clogged with particulates if a quick transition to pure biodiesel is made. Therefore, it is recommended to change the fuel filters on engines and heaters shortly after first switching to a biodiesel blend.

Distribution

Since the passage of the Energy Policy Act of 2005, biodiesel use has been increasing in the United States. In the UK, the Renewable Transport Fuel Obligation obliges suppliers to include 5% renewable fuel in all transport fuel sold in the UK by 2010. For road diesel, this effectively means 5% biodiesel (B5).

Vehicular use and Manufacturer Acceptance

In 2005, Chrysler (then part of DaimlerChrysler) released the Jeep Liberty CRD diesels from the factory into the American market with 5% biodiesel blends, indicating at least partial acceptance of biodiesel as an acceptable diesel fuel additive. In 2007, DaimlerChrysler indicated its intention to increase warranty coverage to 20% biodiesel blends if biofuel quality in the United States can be standardized.

The Volkswagen Group has released a statement indicating that several of its vehicles are compatible with B5 and B100 made from rape seed oil and compatible with the EN 14214 standard. The use of the specified biodiesel type in its cars will not void any warranty.

Mercedes Benz does not allow diesel fuels containing greater than 5% biodiesel (B5) due to concerns about "production shortcomings". Any damages caused by the use of such non-approved fuels will not be covered by the Mercedes-Benz Limited Warranty.

Starting in 2004, the city of Halifax, Nova Scotia decided to update its bus system to allow the fleet of city buses to run entirely on a fish-oil based biodiesel. This caused the city some initial mechanical issues, but after several years of refining, the entire fleet had successfully been converted.

In 2007, McDonalds of UK announced it would start producing biodiesel from the waste oil byproduct of its restaurants. This fuel would be used to run its fleet.

The 2014 Chevy Cruze Clean Turbo Diesel, direct from the factory, will be rated for up to B20 (blend of 20% biodiesel / 80% regular diesel) biodiesel compatibility

Railway Usage

Biodiesel locomotive and its external fuel tank at Mount Washington Cog Railway

British train operating company Virgin Trains claimed to have run the UK's first "biodiesel train", which was converted to run on 80% petrodiesel and 20% biodiesel.

The Royal Train on 15 September 2007 completed its first ever journey run on 100% biodiesel fuel supplied by Green Fuels Ltd. His Royal Highness, The Prince of Wales, and Green Fuels managing director, James Hygate, were the first passengers on a train fueled entirely by biodiesel fuel. Since 2007, the Royal Train has operated successfully on B100 (100% biodiesel).

Similarly, a state-owned short-line railroad in eastern Washington ran a test of a 25% biodiesel / 75% petrodiesel blend during the summer of 2008, purchasing fuel from a biodiesel producer sited along the railroad tracks. The train will be powered by biodiesel made in part from canola grown in agricultural regions through which the short line runs.

Also in 2007, Disneyland began running the park trains on B98 (98% biodiesel). The program was discontinued in 2008 due to storage issues, but in January 2009, it was announced that the park would then be running all trains on biodiesel manufactured from its own used cooking oils. This is a change from running the trains on soy-based biodiesel.

In 2007, the historic Mt. Washington Cog Railway added the first biodiesel locomotive to its all-steam locomotive fleet. The fleet has climbed up the western slopes of Mount Washington in New Hampshire since 1868 with a peak vertical climb of 37.4 degrees.

On 8 July 2014, the then Indian Railway Minister D.V. Sadananda Gowda announced in Railway Budget that 5% bio-diesel will be used in Indian Railways' Diesel Engines.

Aircraft Use

A test flight has been performed by a Czech jet aircraft completely powered on biodiesel. Other recent jet flights using biofuel, however, have been using other types of renewable fuels.

On November 7, 2011 United Airlines flew the world's first commercial aviation flight on a microbially derived biofuel using Solajet™, Solazyme's algae-derived renewable jet fuel. The Eco-skies Boeing 737-800 plane was fueled with 40 percent Solajet and 60 percent petroleum-derived jet fuel. The commercial Eco-skies flight 1403 departed from Houston's IAH airport at 10:30 and landed at Chicago's ORD airport at 13:03.

As a Heating Oil

Biodiesel can also be used as a heating fuel in domestic and commercial boilers, a mix of heating oil and biofuel which is standardized and taxed slightly differently from diesel fuel used for transportation. Bioheat® fuel is a proprietary blend of biodiesel and traditional heating oil. Bioheat® is a registered trademark of the National Biodiesel Board [NBB] and the National Oilheat Research Alliance [NORA] in the U.S., and Columbia Fuels in Canada). Heating biodiesel is available in various blends. ASTM 396 recognizes blends of up to 5 percent biodiesel as equivalent to pure petroleum heating oil. Blends of higher levels of up to 20% biofuel are used by many consumers. Research is underway to determine whether such blends affect performance.

Older furnaces may contain rubber parts that would be affected by biodiesel's solvent properties, but can otherwise burn biodiesel without any conversion required. Care must be taken, however, given that varnishes left behind by petrodiesel will be released and can clog pipes- fuel filtering and prompt filter replacement is required. Another approach is to start using biodiesel as a blend, and decreasing the petroleum proportion over time can allow the varnishes to come off more gradually and be less likely to clog. Thanks to its strong solvent properties, however, the furnace is cleaned out and generally becomes more efficient. A technical research paper describes laboratory research and field trials project using pure biodiesel and biodiesel blends as a heating fuel in oil-fired boilers. During the Biodiesel Expo 2006 in the UK, Andrew J. Robertson presented his biodiesel heating oil research from his technical paper and suggested B20 biodiesel could reduce UK household CO_2 emissions by 1.5 million tons per year.

A law passed under Massachusetts Governor Deval Patrick requires all home heating diesel in that state to be 2% biofuel by July 1, 2010, and 5% biofuel by 2013. New York City has passed a similar law.

Cleaning Oil Spills

With 80-90% of oil spill costs invested in shoreline cleanup, there is a search for more efficient and cost-effective methods to extract oil spills from the shorelines. Biodiesel has displayed its capacity to significantly dissolve crude oil, depending on the source of the fatty acids. In a laboratory setting, oiled sediments that simulated polluted shorelines were sprayed with a single coat of biodiesel and exposed to simulated tides. Biodiesel is an effective solvent to oil due to its methyl ester component, which considerably lowers the viscosity of the crude oil. Additionally, it has a higher buoyancy than crude oil, which later aids in its removal. As a result, 80% of oil was removed from cobble and fine sand, 50% in coarse sand, and 30% in gravel. Once the oil is liberated from the shoreline, the oil-biodiesel mixture is manually removed from the water surface with skimmers. Any remaining mixture is easily broken down due to the high biodegradability of biodiesel, and the increased surface area exposure of the mixture.

Biodiesel in Generators

Biodiesel is also used in rental generators

In 2001, UC Riverside installed a 6-megawatt backup power system that is entirely fueled by biodiesel. Backup diesel-fueled generators allow companies to avoid damaging blackouts of critical operations at the expense of high pollution and emission rates. By using B100, these generators were able to essentially eliminate the byproducts that result in smog, ozone, and sulfur emissions. The use of these generators in residential areas around schools, hospitals, and the general public result in substantial reductions in poisonous carbon monoxide and particulate matter.

Historical Background

Transesterification of a vegetable oil was conducted as early as 1853 by Patrick Duffy, four decades before the first diesel engine became functional. Rudolf Diesel's prime model, a single 10 ft (3.0 m) iron cylinder with a flywheel at its base, ran on its own power for the first time in Augsburg, Germany, on 10 August 1893 running on nothing but peanut oil. In remembrance of this event, 10 August has been declared "International Biodiesel Day".

It is often reported that Diesel designed his engine to run on peanut oil, but this is not the case. Diesel stated in his published papers, "at the Paris Exhibition in 1900 (*Exposition Universelle*) there was shown by the Otto Company a small Diesel engine, which, at the request of the French government ran on arachide (earth-nut or pea-nut) oil, and worked so smoothly that only a few people were aware of it. The engine was constructed for using mineral oil, and was then worked on vegetable oil without any alterations being made. The French Government at the time thought of testing the applicability to power production of the Arachide, or earth-nut, which grows in considerable quantities in their African colonies, and can easily be cultivated there." Diesel himself later conducted related tests and appeared supportive of the idea. In a 1912 speech Diesel said, "the use of vegetable oils for engine fuels may seem insignificant today but such oils may become, in the course of time, as important as petroleum and the coal-tar products of the present time."

Rudolf Diesel

Despite the widespread use of petroleum-derived diesel fuels, interest in vegetable oils as fuels for internal combustion engines was reported in several countries during the 1920s and 1930s and later during World War II. Belgium, France, Italy, the United Kingdom, Portugal, Germany, Brazil, Argentina, Japan and China were reported to have tested and used vegetable oils as diesel fuels during this time. Some operational problems were reported due to the high viscosity of vegetable oils compared to petroleum diesel fuel, which results in poor atomization of the fuel in the fuel spray and often leads to deposits and coking of the injectors, combustion chamber and valves. Attempts to overcome these problems included heating of the vegetable oil, blending it with petroleum-derived diesel fuel or ethanol, pyrolysis and cracking of the oils.

On 31 August 1937, G. Chavanne of the University of Brussels (Belgium) was granted a patent for a "Procedure for the transformation of vegetable oils for their uses as fuels" (fr. "*Procédé de Transformation d'Huiles Végétales en Vue de Leur Utilisation comme Carburants*") Belgian Patent 422,877. This patent described the alcoholysis (often referred to as transesterification) of vegetable oils using ethanol (and mentions methanol) in order to separate the fatty acids from the glycerol by replacing the glycerol with short linear alcohols. This appears to be the first account of the production of what is known as "biodiesel" today.

More recently, in 1977, Brazilian scientist Expedito Parente invented and submitted for patent, the first industrial process for the production of biodiesel. This process is classified as biodiesel by international norms, conferring a "standardized identity and quality. No other proposed biofuel has been validated by the motor industry." As of 2010, Parente's company Tecbio is working with Boeing and NASA to certify bioquerosene (bio-kerosene), another product produced and patented by the Brazilian scientist.

Research into the use of transesterified sunflower oil, and refining it to diesel fuel standards, was initiated in South Africa in 1979. By 1983, the process for producing fuel-quality, engine-tested biodiesel was completed and published internationally. An Austrian company, Gaskoks, obtained the technology from the South African Agricultural Engineers; the company erected the first biodiesel pilot plant in November 1987, and the first industrial-scale plant in April 1989 (with a capacity of 30,000 tons of rapeseed per annum).

Throughout the 1990s, plants were opened in many European countries, including the Czech Republic, Germany and Sweden. France launched local production of biodiesel fuel (referred to as *diester*) from rapeseed oil, which is mixed into regular diesel fuel at a level of 5%, and into the diesel fuel used by some captive fleets (e.g. public transportation) at a level of 30%. Renault, Peugeot and other manufacturers have certified truck engines for use with up to that level of partial biodiesel; experiments with 50% biodiesel are underway. During the same period, nations in other parts of the world also saw local production of biodiesel starting up: by 1998, the Austrian Biofuels Institute had identified 21 countries with commercial biodiesel projects. 100% biodiesel is now available at many normal service stations across Europe.

Properties

Biodiesel has promising lubricating properties and cetane ratings compared to low sulfur diesel fuels. Depending on the engine, this might include high pressure injection pumps, pump injectors (also called *unit injectors*) and fuel injectors.

Older diesel Mercedes are popular for running on biodiesel.

The calorific value of biodiesel is about 37.27 MJ/kg. This is 9% lower than regular Number 2 petrodiesel. Variations in biodiesel energy density is more dependent on the feedstock used than the production process. Still, these variations are less than for petrodiesel. It has been claimed biodiesel gives better lubricity and more complete combustion thus increasing the engine energy output and partially compensating for the higher energy density of petrodiesel.

The color of biodiesel ranges from golden to dark brown, depending on the production method. It is slightly miscible with water, has a high boiling point and low vapor pressure. *The flash point of biodiesel (>130 °C, >266 °F) is significantly higher than that of petroleum diesel (64 °C, 147 °F) or gasoline (−45 °C, -52 °F). Biodiesel has a density of ~ 0.88 g/cm³, higher than petrodiesel (~ 0.85 g/cm³).

Biodiesel contains virtually no sulfur, and it is often used as an additive to Ultra-Low Sulfur Diesel (ULSD) fuel to aid with lubrication, as the sulfur compounds in petrodiesel provide much of the lubricity.

Fuel Efficiency

The power output of biodiesel depends on its blend, quality, and load conditions under which the fuel is burnt. The thermal efficiency for example of B100 as compared to B20 will vary due to the differing energy content of the various blends. Thermal efficiency of a fuel is based in part on fuel characteristics such as: viscosity, specific density, and flash point; these characteristics will change as the blends as well as the quality of biodiesel varies. The American Society for Testing and Materials has set standards in order to judge the quality of a given fuel sample.

Regarding brake thermal efficiency one study found that B40 was superior to traditional counterpart at higher compression ratios (this higher brake thermal efficiency was recorded at compression ratios of 21:1). It was noted that, as the compression ratios increased, the efficiency of all fuel types - as well as blends being tested - increased; though it was found that a blend of B40 was the most economical at a compression ratio of 21:1 over all other blends. The study implied that this increase in efficiency was due to fuel density, viscosity, and heating values of the fuels.

Combustion

Fuel systems on the modern diesel engine were not designed to accommodate biodiesel, while many heavy duty engines are able to run with biodiesel blends e.g. B20. Traditional direct injection fuel systems operate at roughly 3,000 psi at the injector tip while the modern common rail fuel system operates upwards of 30,000 PSI at the injector tip. Components are designed to operate at a great temperature range, from below freezing to over 1,000 degrees Fahrenheit. Diesel fuel is expected to burn efficiently and produce as few emissions as possible. As emission standards are being introduced to diesel engines the need to control harmful emissions is being designed into the parameters of diesel engine fuel systems. The traditional inline injection system is more forgiving to poorer quality fuels as opposed to the common rail fuel system. The higher pressures and tighter tolerances of the common rail system allows for greater control over atomization and injection timing. This control of atomization as well as combustion allows for greater efficiency of modern diesel engines as well as greater control over emissions. Components within a diesel fuel system interact with the fuel in a way to ensure efficient operation of the fuel system and so the engine. If an out-of-specification fuel is introduced to a system that has specific parameters of operation, then the integrity of the overall fuel system may be compromised. Some of these parameters such as spray pattern and atomization are directly related to injection timing.

One study found that during atomization biodiesel and its blends produced droplets were greater in diameter than the droplets produced by traditional petrodiesel. The smaller droplets were attributed to the lower viscosity and surface tension of traditional petrol. It was found that droplets at the periphery of the spray pattern were larger in diameter than the droplets at the center this was attributed to the faster pressure drop at the edge of the spray pattern; there was a proportional relationship between the droplet size and the distance from the injector tip. It was found that B100 had the greatest spray penetration, this was attributed to the greater density of B100. Having a greater droplet size can lead to; inefficiencies in the combustion, increased emissions, and decreased horse power. In another study it was found that there is a short injection delay when injecting biodiesel. This injection delay was attributed to the greater viscosity of Biodiesel. It was noted that the higher viscosity and the greater cetane rating of biodiesel over traditional petrodiesel lead to poor atomization, as well as mixture penetration with air during the ignition delay period. Another study noted that this ignition delay may aid in a decrease of NOx emission.

Emissions

Emissions are inherent to the combustion of diesel fuels that are regulated by the U.S. Environmental Protection Agency (E.P.A.). As these emissions are a byproduct of the combustion process, in order to ensure E.P.A. compliance a fuel system must be capable of controlling the combustion of fuels as well as the mitigation of emissions. There are a number of new technologies being phased in to control the production of diesel emissions. The exhaust gas recirculation system, E.G.R., and the diesel particulate filter, D.P.F., are both designed to mitigate the production of harmful emissions.

A study performed by the Chonbuk National University concluded that a B30 biodiesel blend reduced carbon monoxide emissions by approximately 83% and particulate matter emissions by roughly 33%. NOx emissions, however, were found to increase without the application of an E.G.R. system. The study also concluded that, with E.G.R, a B20 biodiesel blend considerably reduced the emissions of the engine. Additionally, analysis by the California Air Resources Board found that biodiesel had the lowest carbon emissions of the fuels tested, those being ultra-low-sulfur diesel, gasoline, corn-based ethanol, compressed natural gas, and five types of biodiesel from varying feedstocks. Their conclusions also showed great variance in carbon emissions of biodiesel based on the feedstock used. Of soy, tallow, canola, corn, and used cooking oil, soy showed the highest carbon emissions, while used cooking oil produced the lowest.

While studying the effect of biodiesel on a D.P.F. it was found that though the presence of sodium and potassium carbonates aided in the catalytic conversion of ash, as the diesel particulates are catalyzed, they may congregate inside the D.P.F. and so interfere with the clearances of the filter. This may cause the filter to clog and interfere with the regeneration process. In a study on the impact of E.G.R. rates with blends of jathropa biodiesel it was shown that there was a decrease in fuel efficiency and torque output due to the use of biodiesel on a diesel engine designed with an E.G.R. system. It was found that CO and CO2 emissions increased with an increase in exhaust gas recirculation but NOx levels decreased. The opacity level of the jathropa blends was in an acceptable range, where traditional diesel was out of acceptable standards. It was shown that a decrease in Nox emissions could be obtained with an E.G.R. system. This study showed an advantage over traditional diesel within a certain operating range of the E.G.R. system. Currently blended biodiesel fuels (B5 and B20) are being used in many heavy-duty vehicles especially transit buses in US cities. Characterization of exhaust emissions showed significant emission reductions compared to regular diesel.

Material Compatibility

- Plastics: High-density polyethylene (HDPE) is compatible but polyvinyl chloride (PVC) is slowly degraded. Polystyrene is dissolved on contact with biodiesel.
- Metals: Biodiesel (like methanol) has an effect on copper-based materials (e.g. brass), and it also affects zinc, tin, lead, and cast iron. Stainless steels (316 and 304) and aluminum are unaffected.
- Rubber: Biodiesel also affects types of natural rubbers found in some older engine components. Studies have also found that fluorinated elastomers (FKM) cured with peroxide and base-metal oxides can be degraded when biodiesel loses its stability caused by oxidation. Commonly used synthetic rubbers FKM-GBL-S and FKM- GF-S found in modern vehicles were found to handle biodiesel in all conditions.

Technical Standards

Biodiesel has a number of standards for its quality including European standard EN 14214, ASTM International D6751, and others.

Low Temperature Gelling

When biodiesel is cooled below a certain point, some of the molecules aggregate and form crystals. The fuel starts to appear cloudy once the crystals become larger than one quarter of the wavelengths of visible light - this is the cloud point (CP). As the fuel is cooled further these crystals become larger. The lowest temperature at which fuel can pass through a 45 micrometre filter is the cold filter plugging point (CFPP). As biodiesel is cooled further it will gel and then solidify. Within Europe, there are differences in the CFPP requirements between countries. This is reflected in the different national standards of those countries. The temperature at which pure (B100) biodiesel starts to gel varies significantly and depends upon the mix of esters and therefore the feedstock oil used to produce the biodiesel. For example, biodiesel produced from low erucic acid varieties of canola seed (RME) starts to gel at approximately −10 °C (14 °F). Biodiesel produced from beef tallow and palm oil tends to gel at around 16 °C (61 °F) and 13 °C (55 °F) respectively. There are a number of commercially available additives that will significantly lower the pour point and cold filter plugging point of pure biodiesel. Winter operation is also possible by blending biodiesel with other fuel oils including #2 low sulfur diesel fuel and #1 diesel / kerosene.

Another approach to facilitate the use of biodiesel in cold conditions is by employing a second fuel tank for biodiesel in addition to the standard diesel fuel tank. The second fuel tank can be insulated and a heating coil using engine coolant is run through the tank. The fuel tanks can be switched over when the fuel is sufficiently warm. A similar method can be used to operate diesel vehicles using straight vegetable oil.

Contamination by Water

Biodiesel may contain small but problematic quantities of water. Although it is only slightly miscible with water it is hygroscopic. One of the reasons biodiesel can absorb water is the persistence of mono and diglycerides left over from an incomplete reaction. These molecules can act as an emulsifier, allowing water to mix with the biodiesel. In addition, there may be water that is residual to processing or resulting from storage tank condensation. The presence of water is a problem because:

- Water reduces the heat of fuel combustion, causing smoke, harder starting, and reduced power.
- Water causes corrosion of fuel system components (pumps, fuel lines, etc.)
- Microbes in water cause the paper-element filters in the system to rot and fail, causing failure of the fuel pump due to ingestion of large particles.
- Water freezes to form ice crystals that provide sites for nucleation, accelerating gelling of the fuel.
- Water causes pitting in pistons.

Previously, the amount of water contaminating biodiesel has been difficult to measure by taking samples, since water and oil separate. However, it is now possible to measure the water content using water-in-oil sensors.

Water contamination is also a potential problem when using certain chemical catalysts involved in the production process, substantially reducing catalytic efficiency of base (high pH) catalysts such as potassium hydroxide. However, the super-critical methanol production methodology, whereby the transesterification process of oil feedstock and methanol is effectuated under high temperature and pressure, has been shown to be largely unaffected by the presence of water contamination during the production phase.

Availability and Prices

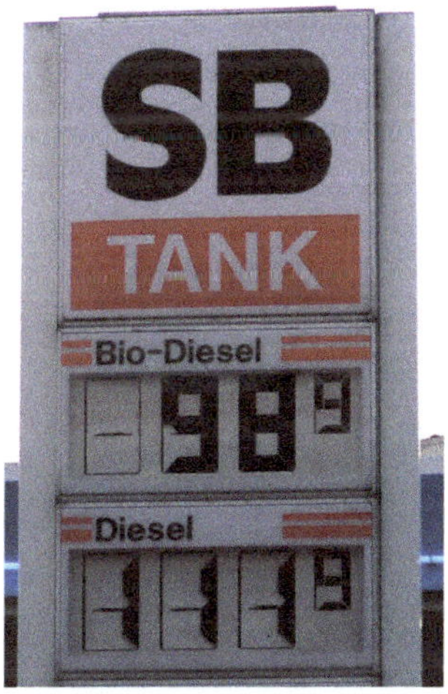

In some countries biodiesel is less expensive than conventional diesel

Global biodiesel production reached 3.8 million tons in 2005. Approximately 85% of biodiesel production came from the European Union.

In 2007, in the United States, average retail (at the pump) prices, including federal and state fuel taxes, of B2/B5 were lower than petroleum diesel by about 12 cents, and B20 blends were the same as petrodiesel. However, as part of a dramatic shift in diesel pricing, by July 2009, the US DOE was reporting average costs of B20 15 cents per gallon higher than petroleum diesel ($2.69/gal vs. $2.54/gal). B99 and B100 generally cost more than petrodiesel except where local governments provide a tax incentive or subsidy.

Production

Biodiesel is commonly produced by the transesterification of the vegetable oil or animal fat feedstock. There are several methods for carrying out this transesterification reaction including the common batch process, supercritical processes, ultrasonic methods, and even microwave methods.

Chemically, transesterified biodiesel comprises a mix of mono-alkyl esters of long chain fatty acids. The most common form uses methanol (converted to sodium methoxide) to produce methyl esters (commonly referred to as Fatty Acid Methyl Ester - FAME) as it is the cheapest alcohol available, though ethanol can be used to produce an ethyl ester (commonly referred to as Fatty Acid Ethyl Ester - FAEE) biodiesel and higher alcohols such as isopropanol and butanol have also been used. Using alcohols of higher molecular weights improves the cold flow properties of the resulting ester, at the cost of a less efficient transesterification reaction. A lipid transesterification production process is used to convert the base oil to the desired esters. Any free fatty acids (FFAs) in the base oil are either converted to soap and removed from the process, or they are esterified (yielding more biodiesel) using an acidic catalyst. After this processing, unlike straight vegetable oil, biodiesel has combustion properties very similar to those of petroleum diesel, and can replace it in most current uses.

The methanol used in most biodiesel production processes is made using fossil fuel inputs. However, there are sources of renewable methanol made using carbon dioxide or biomass as feedstock, making their production processes free of fossil fuels.

A by-product of the transesterification process is the production of glycerol. For every 1 tonne of biodiesel that is manufactured, 100 kg of glycerol are produced. Originally, there was a valuable market for the glycerol, which assisted the economics of the process as a whole. However, with the increase in global biodiesel production, the market price for this crude glycerol (containing 20% water and catalyst residues) has crashed. Research is being conducted globally to use this glycerol as a chemical building block. One initiative in the UK is The Glycerol Challenge.

Usually this crude glycerol has to be purified, typically by performing vacuum distillation. This is rather energy intensive. The refined glycerol (98%+ purity) can then be utilised directly, or converted into other products. The following announcements were made in 2007: A joint venture of Ashland Inc. and Cargill announced plans to make propylene glycol in Europe from glycerol and Dow Chemical announced similar plans for North America. Dow also plans to build a plant in China to make epichlorhydrin from glycerol. Epichlorhydrin is a raw material for epoxy resins.

Production Levels

In 2007, biodiesel production capacity was growing rapidly, with an average annual growth rate from 2002-06 of over 40%. For the year 2006, the latest for which actual production figures could be obtained, total world biodiesel production was about 5-6 million tonnes, with 4.9 million tonnes processed in Europe (of which 2.7 million tonnes was from Germany) and most of the rest from the USA. In 2008 production in Europe alone had risen to 7.8 million tonnes. In July 2009, a duty was added to American imported biodiesel in the European Union in order to balance the competition from European, especially German producers. The capacity for 2008 in Europe totalled

16 million tonnes. This compares with a total demand for diesel in the US and Europe of approximately 490 million tonnes (147 billion gallons). Total world production of vegetable oil for all purposes in 2005/06 was about 110 million tonnes, with about 34 million tonnes each of palm oil and soybean oil.

US biodiesel production in 2011 brought the industry to a new milestone. Under the EPA Renewable Fuel Standard, targets have been implemented for the biodiesel production plants in order to monitor and document production levels in comparison to total demand. According to the year-end data released by the EPA, biodiesel production in 2011 reached more than 1 billion gallons. This production number far exceeded the 800 million gallon target set by the EPA. The projected production for 2020 is nearly 12 billion gallons.

Biodiesel Feedstocks

Plant oils

Soybeans are used as a source of biodiesel

A variety of oils can be used to produce biodiesel. These include:

- Virgin oil feedstock – rapeseed and soybean oils are most commonly used, soybean oil accounting for about half of U.S. production. It also can be obtained from Pongamia, field pennycress and jatropha and other crops such as mustard, jojoba, flax, sunflower, palm oil, coconut and hemp;
- Waste vegetable oil (WVO);
- Animal fats including tallow, lard, yellow grease, chicken fat, and the by-products of the production of Omega-3 fatty acids from fish oil.

- Algae, which can be grown using waste materials such as sewage and without displacing land currently used for food production.
- Oil from halophytes such as *Salicornia bigelovii*, which can be grown using saltwater in coastal areas where conventional crops cannot be grown, with yields equal to the yields of soybeans and other oilseeds grown using freshwater irrigation
- Sewage Sludge - The sewage-to-biofuel field is attracting interest from major companies like Waste Management and startups like InfoSpi, which are betting that renewable sewage biodiesel can become competitive with petroleum diesel on price.

Many advocates suggest that waste vegetable oil is the best source of oil to produce biodiesel, but since the available supply is drastically less than the amount of petroleum-based fuel that is burned for transportation and home heating in the world, this local solution could not scale to the current rate of consumption.

Animal fats are a by-product of meat production and cooking. Although it would not be efficient to raise animals (or catch fish) simply for their fat, use of the by-product adds value to the livestock industry (hogs, cattle, poultry). Today, multi-feedstock biodiesel facilities are producing high quality animal-fat based biodiesel. Currently, a 5-million dollar plant is being built in the USA, with the intent of producing 11.4 million litres (3 million gallons) biodiesel from some of the estimated 1 billion kg (2.2 billion pounds) of chicken fat produced annually at the local Tyson poultry plant. Similarly, some small-scale biodiesel factories use waste fish oil as feedstock. An EU-funded project (ENERFISH) suggests that at a Vietnamese plant to produce biodiesel from catfish (basa, also known as pangasius), an output of 13 tons/day of biodiesel can be produced from 81 tons of fish waste (in turn resulting from 130 tons of fish). This project utilises the biodiesel to fuel a CHP unit in the fish processing plant, mainly to power the fish freezing plant.

Quantity of Feedstocks Required

Current worldwide production of vegetable oil and animal fat is not sufficient to replace liquid fossil fuel use. Furthermore, some object to the vast amount of farming and the resulting fertilization, pesticide use, and land use conversion that would be needed to produce the additional vegetable oil. The estimated transportation diesel fuel and home heating oil used in the United States is about 160 million tons (350 billion pounds) according to the Energy Information Administration, US Department of Energy. In the United States, estimated production of vegetable oil for all uses is about 11 million tons (24 billion pounds) and estimated production of animal fat is 5.3 million tonnes (12 billion pounds).

If the entire arable land area of the USA (470 million acres, or 1.9 million square kilometers) were devoted to biodiesel production from soy, this would just about provide the 160 million tonnes required (assuming an optimistic 98 US gal/acre of biodiesel). This land area could in principle be reduced significantly using algae, if the obstacles can be overcome. The US DOE estimates that if algae fuel replaced all the petroleum fuel in the United States, it would require 15,000 square miles (39,000 square kilometers), which is a few thousand square miles larger than Maryland, or 30% greater than the area of Belgium, assuming a yield of 140 tonnes/hectare (15,000 US gal/acre). Given a more realistic yield of 36 tonnes/hectare (3834 US gal/acre) the area required is about 152,000 square kilometers, or roughly equal to that of the state of Georgia or of England and Wales. The advantages of algae are that it can be grown on non-arable land such as deserts or in marine environments, and the potential oil yields are much higher than from plants.

Yield

Feedstock yield efficiency per unit area affects the feasibility of ramping up production to the huge industrial levels required to power a significant percentage of vehicles.

Some Typical Yields

Crop	**Yield**	
	L/ha	**US gal/acre**
Palm oil	4752	508
Coconut	2151	230
Cyperus esculentus	1628	174
Rapeseed	954	102
Soy (Indiana)	554-922	59.2-98.6
Chinese tallow	907	97
Peanut	842	90
Sunflower	767	82
Hemp	242	26

1. *"Biofuels: some numbers". Grist.org. Retrieved 2010-03-15.*
2. Makareviciene et al., "Opportunities for the use of chufa sedge in biodiesel production", Industrial Crops and Products, 50 (2013) p. 635, table 2.
3. Klass, Donald, "Biomass for Renewable Energy, Fuels, and Chemicals", page 341. Academic Press, 1998.
4. Kitani, Osamu, "Volume V: Energy and Biomass Engineering, CIGR Handbook of Agricultural Engineering", Amer Society of Agricultural, 1999.

Algae fuel yields have not yet been accurately determined, but DOE is reported as saying that algae yield 30 times more energy per acre than land crops such as soybeans. Yields of 36 tonnes/hectare are considered practical by Ami Ben-Amotz of the Institute of Oceanography in Haifa, who has been farming Algae commercially for over 20 years.

Jatropha has been cited as a high-yield source of biodiesel but yields are highly dependent on climatic and soil conditions. The estimates at the low end put the yield at about 200 US gal/acre (1.5-2 tonnes per hectare) per crop; in more favorable climates two or more crops per year have been achieved. It is grown in the Philippines, Mali and India, is drought-resistant, and can share space with other cash crops such as coffee, sugar, fruits and vegetables. It is well-suited to semi-arid lands and can contribute to slow down desertification, according to its advocates.

Efficiency and Economic Arguments

Pure biodiesel (B-100) made from soybeans

According to a study by Drs. Van Dyne and Raymer for the Tennessee Valley Authority, the average US farm consumes fuel at the rate of 82 litres per hectare (8.75 US gal/acre) of land to produce one crop. However, average crops of rapeseed produce oil at an average rate of 1,029 L/ha (110 US gal/acre), and high-yield rapeseed fields produce about 1,356 L/ha (145 US gal/acre). The ratio of input to output in these cases is roughly 1:12.5 and 1:16.5. Photosynthesis is known to have an efficiency rate of about 3-6% of total solar radiation and if the entire mass of a crop is utilized for energy production, the overall efficiency of this chain is currently about 1% While this may compare unfavorably to solar cells combined with an electric drive train, biodiesel is less costly to deploy (solar cells cost approximately US$250 per square meter) and transport (electric vehicles require batteries which currently have a much lower energy density than liquid fuels). A 2005 study found that biodiesel production using soybeans required 27% more fossil energy than the biodiesel produced and 118% more energy using sunflowers.

However, these statistics by themselves are not enough to show whether such a change makes economic sense. Additional factors must be taken into account, such as: the fuel equivalent of the energy required for processing, the yield of fuel from raw oil, the return on cultivating food, the effect biodiesel will have on food prices and the relative cost of biodiesel versus petrodiesel, water pollution from farm run-off, soil depletion, and the externalized costs of political and military interference in oil-producing countries intended to control the price of petrodiesel.

The debate over the energy balance of biodiesel is ongoing. Transitioning fully to biofuels could require immense tracts of land if traditional food crops are used (although non food crops can be utilized). The problem would be especially severe for nations with large economies, since energy consumption scales with economic output.

If using only traditional food plants, most such nations do not have sufficient arable land to produce biofuel for the nation's vehicles. Nations with smaller economies (hence less energy consumption) and more arable land may be in better situations, although many regions cannot afford to divert land away from food production.

For third world countries, biodiesel sources that use marginal land could make more sense; e.g., pongam oiltree nuts grown along roads or jatropha grown along rail lines.

In tropical regions, such as Malaysia and Indonesia, plants that produce palm oil are being planted at a rapid pace to supply growing biodiesel demand in Europe and other markets. Scientists have shown that the removal of rainforest for palm plantations is not ecologically sound since the expansion of oil palm plantations poses a threat to natural rainforest and biodiversity.

It has been estimated in Germany that palm oil biodiesel has less than one third of the production costs of rapeseed biodiesel. The direct source of the energy content of biodiesel is solar energy captured by plants during photosynthesis. Regarding the positive energy balance of biodiesel:

> When straw was left in the field, biodiesel production was strongly energy positive, yielding 1 GJ biodiesel for every 0.561 GJ of energy input (a yield/cost ratio of 1.78).
>
> When straw was burned as fuel and oilseed rapemeal was used as a fertilizer, the yield/cost ratio for biodiesel production was even better (3.71). In other words, for every unit of energy input to produce biodiesel, the output was 3.71 units (the difference of 2.71 units would be from solar energy).

Economic Impact

Multiple economic studies have been performed regarding the economic impact of biodiesel production. One study, commissioned by the National Biodiesel Board, reported the 2011 production of biodiesel supported 39,027 jobs and more than $2.1 billion in household income. The growth in biodiesel also helps significantly increase GDP. In 2011, biodiesel created more than $3 billion in GDP. Judging by the continued growth in the Renewable Fuel Standard and the extension of the biodiesel tax incentive, the number of jobs can increase to 50,725, $2.7 billion in income, and reaching $5 billion in GDP by 2012 and 2013.

Energy Security

One of the main drivers for adoption of biodiesel is energy security. This means that a nation's dependence on oil is reduced, and substituted with use of locally available sources, such as coal, gas, or renewable sources. Thus a country can benefit from adoption of biofuels, without a reduction in greenhouse gas emissions. While the total energy balance is debated, it is clear that the dependence on oil is reduced. One example is the energy used to manufacture fertilizers, which could come from a variety of sources other than petroleum. The US National Renewable Energy Laboratory (NREL) states that energy security is the number one driving force behind the US biofuels programme, and a White House "Energy Security for the 21st Century" paper makes it clear that energy security is a major reason for promoting biodiesel. The former EU commission president, Jose Manuel Barroso, speaking at a recent EU biofuels conference, stressed that properly managed biofuels have the potential to reinforce the EU's security of supply through diversification of energy sources.

Global Biofuel Policies

Many countries around the world are involved in the growing use and production of biofuels, such as biodiesel, as an alternative energy source to fossil fuels and oil. To foster the biofuel industry, governments have implemented legislations and laws as incentives to reduce oil dependency and to increase the use of renewable energies. Many countries have their own independent policies regarding the taxation and rebate of biodiesel use, import, and production.

Canada

It was required by the Canadian Environmental Protection Act Bill C-33 that by the year 2010, gasoline contained 5% renewable content and that by 2013, diesel and heating oil contained 2% renewable content. The EcoENERGY for Biofuels Program subsidized the production of biodiesel, among other biofuels, via an incentive rate of CAN$0.20 per liter from 2008 to 2010. A decrease of $0.04 will be applied every year following, until the incentive rate reaches $0.06 in 2016. Individual provinces also have specific legislative measures in regards to biofuel use and production.

United States

The Volumetric Ethanol Excise Tax Credit (VEETC) was the main source of financial support for biofuels, but was scheduled to expire in 2010. Through this act, biodiesel production guaranteed a tax credit of US$1 per gallon produced from virgin oils, and $0.50 per gallon made from recycled oils. Currently soybean oil is being used to produce soybean biodiesel for many commercial purposes such as blending fuel for transportation sectors.

European Union

The European Union is the greatest producer of biodiesel, with France and Germany being the top producers. To increase the use of biodiesel, there exist policies requiring the blending of biodiesel into fuels, including penalties if those rates are not reached. In France, the goal was to reach 10% integration but plans for that stopped in 2010. As an incentive for the European Union countries to continue the production of the biofuel, there are tax rebates for specific quotas of biofuel produced. In Germany, the minimum percentage of biodiesel in transport diesel is set at 7% so called "B7".

Environmental Effects

The surge of interest in biodiesels has highlighted a number of environmental effects associated with its use. These potentially include reductions in greenhouse gas emissions, deforestation, pollution and the rate of biodegradation.

According to the EPA's Renewable Fuel Standards Program Regulatory Impact Analysis, released in February 2010, biodiesel from soy oil results, on average, in a 57% reduction in greenhouse gases compared to petroleum diesel, and biodiesel produced from waste grease results in an 86% reduction.

However, environmental organizations, for example, Rainforest Rescue and Greenpeace, criticize the cultivation of plants used for biodiesel production, e.g., oil palms, soybeans and sugar cane. They say the deforestation of rainforests exacerbates climate change and that sensitive ecosystems are destroyed to clear land for oil palm, soybean and sugar cane plantations. Moreover, that biofuels contribute to world hunger, seeing as arable land is no longer used for growing foods. The Environmental Protection Agency(EPA) published data in January 2012, showing that biofuels made from palm oil won't count towards the nation's renewable fuels mandate as they are not climate-friendly. Environmentalists welcome the conclusion because the growth of oil palm plantations has driven tropical deforestation, for example, in Indonesia and Malaysia.

Food, Land and Water vs. Fuel

In some poor countries the rising price of vegetable oil is causing problems. Some propose that fuel only be made from non-edible vegetable oils such as camelina, jatropha or seashore mallow which can thrive on marginal agricultural land where many trees and crops will not grow, or would produce only low yields.

Others argue that the problem is more fundamental. Farmers may switch from producing food crops to producing biofuel crops to make more money, even if the new crops are not edible. The law of supply and demand predicts that if fewer farmers are producing food the price of food will rise. It may take some time, as farmers can take some time to change which things they are growing, but increasing demand for first generation biofuels is likely to result in price increases for many kinds of food. Some have pointed out that there are poor farmers and poor countries who are making more money because of the higher price of vegetable oil.

Biodiesel from sea algae would not necessarily displace terrestrial land currently used for food production and new algaculture jobs could be created.

Current Research

There is ongoing research into finding more suitable crops and improving oil yield. Other sources are possible including human fecal matter, with Ghana building its first "fecal sludge-fed biodiesel plant." Using the current yields, vast amounts of land and fresh water would be needed to produce enough oil to completely replace fossil fuel usage. It would require twice the land area of the US to be devoted to soybean production, or two-thirds to be devoted to rapeseed production, to meet current US heating and transportation needs.

Specially bred mustard varieties can produce reasonably high oil yields and are very useful in crop rotation with cereals, and have the added benefit that the meal leftover after the oil has been pressed out can act as an effective and biodegradable pesticide.

The NFESC, with Santa Barbara-based Biodiesel Industries is working to develop biodiesel technologies for the US navy and military, one of the largest diesel fuel users in the world.

A group of Spanish developers working for a company called Ecofasa announced a new biofuel made from trash. The fuel is created from general urban waste which is treated by bacteria to produce fatty acids, which can be used to make biodiesel.

Another approach that does not require the use of chemical for the production involves the use of genetically modified microbes.

Algal Biodiesel

From 1978 to 1996, the U.S. NREL experimented with using algae as a biodiesel source in the "Aquatic Species Program". A self-published article by Michael Briggs, at the UNH Biodiesel Group, offers estimates for the realistic replacement of all vehicular fuel with biodiesel by utilizing algae that have a natural oil content greater than 50%, which Briggs suggests can be grown on algae ponds at wastewater treatment plants. This oil-rich algae can then be extracted from the system and processed into biodiesel, with the dried remainder further reprocessed to create ethanol.

The production of algae to harvest oil for biodiesel has not yet been undertaken on a commercial scale, but feasibility studies have been conducted to arrive at the above yield estimate. In addition to its projected high yield, algaculture — unlike crop-based biofuels — does not entail a decrease in food production, since it requires neither farmland nor fresh water. Many companies are pursuing algae bio-reactors for various purposes, including scaling up biodiesel production to commercial levels.

Prof. Rodrigo E. Teixeira from the University of Alabama in Huntsville demonstrated the extraction of biodiesel lipids from wet algae using a simple and economical reaction in ionic liquids.

Pongamia

Millettia pinnata, also known as the Pongam Oiltree or Pongamia, is a leguminous, oilseed-bearing tree that has been identified as a candidate for non-edible vegetable oil production.

Pongamia plantations for biodiesel production have a two-fold environmental benefit. The trees both store carbon and produce fuel oil. Pongamia grows on marginal land not fit for food crops and does not require nitrate fertilizers. The oil producing tree has the highest yield of oil producing plant (approximately 40% by weight of the seed is oil) while growing in malnourished soils with high levels of salt. It is becoming a main focus in a number of biodiesel research organizations. The main advantages of Pongamia are a higher recovery and quality of oil than other crops and no direct competition with food crops. However, growth on marginal land can lead to lower oil yields which could cause competition with food crops for better soil.

Jatropha

Jatropha Biodiesel from DRDO, India.

Several groups in various sectors are conducting research on Jatropha curcas, a poisonous shrub-like tree that produces seeds considered by many to be a viable source of biodiesel feedstock oil. Much of this research focuses on improving the overall per acre oil yield of Jatropha through advancements in genetics, soil science, and horticultural practices.

SG Biofuels, a San Diego-based Jatropha developer, has used molecular breeding and biotechnology to produce elite hybrid seeds of Jatropha that show significant yield improvements over first generation varieties. SG Biofuels also claims that additional benefits have arisen from such strains, including improved flowering synchronicity, higher resistance to pests and disease, and increased cold weather tolerance.

Plant Research International, a department of the Wageningen University and Research Centre in the Netherlands, maintains an ongoing Jatropha Evaluation Project (JEP) that examines the feasibility of large scale Jatropha cultivation through field and laboratory experiments.

The Center for Sustainable Energy Farming (CfSEF) is a Los Angeles-based non-profit research organization dedicated to Jatropha research in the areas of plant science, agronomy, and horticulture. Successful exploration of these disciplines is projected to increase Jatropha farm production yields by 200-300% in the next ten years.

Fungi

A group at the Russian Academy of Sciences in Moscow published a paper in September 2008, stating that they had isolated large amounts of lipids from single-celled fungi and turned it into biodiesel in an economically efficient manner. More research on this fungal species; *Cunninghamella japonica*, and others, is likely to appear in the near future.

The recent discovery of a variant of the fungus *Gliocladium roseum* points toward the production of so-called myco-diesel from cellulose. This organism was recently discovered in the rainforests of northern Patagonia and has the unique capability of converting cellulose into medium length hydrocarbons typically found in diesel fuel.

Biodiesel from Used Coffee Grounds

Researchers at the University of Nevada, Reno, have successfully produced biodiesel from oil derived from used coffee grounds. Their analysis of the used grounds showed a 10% to 15% oil content (by weight). Once the oil was extracted, it underwent conventional processing into biodiesel. It is estimated that finished biodiesel could be produced for about one US dollar per gallon. Further, it was reported that "the technique is not difficult" and that "there is so much coffee around that several hundred million gallons of biodiesel could potentially be made annually." However, even if all the coffee grounds in the world were used to make fuel, the amount produced would be less than 1 percent of the diesel used in the United States annually. "It won't solve the world's energy problem," Dr. Misra said of his work.

Exotic Sources

Recently, alligator fat was identified as a source to produce biodiesel. Every year, about 15 million pounds of alligator fat are disposed of in landfills as a waste byproduct of the alligator meat and skin industry. Studies have shown that biodiesel produced from alligator fat is similar in composition to biodiesel created from soybeans, and is cheaper to refine since it is primarily a waste product.

Biodiesel to Hydrogen-cell Power

A microreactor has been developed to convert biodiesel into hydrogen steam to power fuel cells.

Steam reforming, also known as fossil fuel reforming is a process which produces hydrogen gas from hydrocarbon fuels, most notably biodiesel due to its efficiency. A **microreactor**, or reformer, is the processing device in which water vapour reacts with the liquid fuel under high temperature and pressure. Under temperatures ranging from 700 – 1100 °C, a nickel-based catalyst enables the production of carbon monoxide and hydrogen:

Hydrocarbon + $H2O \rightleftharpoons CO + 3\ H2$ *(Highly endothermic)*

Furthermore, a higher yield of hydrogen gas can be harnessed by further oxidizing carbon monoxide to produce more hydrogen and carbon dioxide:

$CO + H2O \rightarrow CO2 + H2$ *(Mildly exothermic)*

Hydrogen Fuel Cells Background Information

Fuel cells operate similar to a battery in that electricity is harnessed from chemical reactions. The difference in fuel cells when compared to batteries is their ability to be powered by the constant flow of hydrogen found in the atmosphere. Furthermore, they produce only water as a by-product, and are virtually silent. The downside of hydrogen powered fuel cells is the high cost and dangers of storing highly combustible hydrogen under pressure.

One way new processors can overcome the dangers of transporting hydrogen is to produce it as necessary. The microreactors can be joined to create a system that heats the hydrocarbon under high pressure to generate hydrogen gas and carbon dioxide, a process called steam reforming. This produces up to 160 gallons of hydrogen/minute and gives the potential of powering hydrogen refueling stations, or even an on-board hydrogen fuel source for hydrogen cell vehicles. Implementation into cars would allow energy-rich fuels, such as biodiesel, to be transferred to kinetic energy while avoiding combustion and pollutant byproducts. The hand-sized square piece of metal contains microscopic channels with catalytic sites, which continuously convert biodiesel, and even its glycerol byproduct, to hydrogen.

Concerns

Engine Wear

Lubricity of fuel plays an important role in wear that occurs in an engine. An engine relies on its fuel to provide lubricity for the metal components that are constantly in contact with each other. Biodiesel is a much better lubricant compared with petroleum diesel due to the presence of esters. Tests have shown that the addition of a small amount of biodiesel to diesel can significantly increase the lubricity of the fuel in short term. However, over a longer period of time (2–4 years), studies show that biodiesel loses its lubricity. This could be because of enhanced corrosion over time due to oxidation of the unsaturated molecules or increased water content in biodiesel from moisture absorption.

Fuel Viscosity

One of the main concerns regarding biodiesel is its viscosity. The viscosity of diesel is 2.5–3.2 cSt at 40 °C and the viscosity of biodiesel made from soybean oil is between 4.2 and 4.6 cSt The viscosity of diesel must be high enough to provide sufficient lubrication for the engine parts but low enough to flow at operational temperature. High viscosity can plug the fuel filter and injection system in engines. Vegetable oil is composed of lipids with long chains of hydrocarbons, to reduce its viscosity the lipids are broken down into smaller molecules of esters. This is done by converting vegetable oil and animal fats into alkyl esters using transesterification to reduce their viscosity Nevertheless, biodiesel viscosity remains higher than that of diesel, and the engine may not be able to use the fuel at low temperatures due to the slow flow through the fuel filter.

Engine Performance

Biodiesel has higher brake-specific fuel consumption compared to diesel, which means more biodiesel fuel consumption is required for the same torque. However, B20 biodiesel blend has been found to provide maximum increase in thermal efficiency, lowest brake-specific energy consumption, and lower harmful emissions. The engine performance depends on the properties of the fuel, as well as on combustion, injector pressure and many other factors. Since there are various blends of biodiesel, that may account for the contradicting reports in regards engine performance.

Biodiesel by Region

This chapter describes the use and availability of biodiesel in various countries around the world.

Argentina

Argentina's biodiesel industries are booming as a result of domestic demand along with strong export markets. Biodiesel production in Argentina grew from 130,000 tons in 2006 to 2.5 millon tons in 2010, expecting to produce 3 million tons by 2011. Argentina is a net exporter of biodiesel, nearly all of which was shipped to Europe, overtaking in production the U.S. this year. Argentina ranks as the world's fourth largest producer (by 2011, after Germany, France and Brazil), due to its rapidly emerging domestic market. Argentina is considering a raise from B7 to B10, increasing consumption to 1.3 million tons per year but negotiations need to happen with the automotive industry first.

The Argentine biodiesel industry is mainly based on the use of soybean as feedstock. Production is geographically concentrated in the provinces of Buenos Aires and Santa Fe. Almost all major players in Argentina's vegetable oil industry own or are constructing biodiesel plants.

Australia

The Fuel Standard (Biodiesel) Determination 2003 was signed by the Minister for the Environment and Heritage on 18 September 2006. The determination sets out the physical and chemical parameters of the Biodiesel standard. It also sets out the associated test methods that the Government will use to determine compliance.

Biodiesel subsidies are to be phased out by 2011, after the passing of the Fuel Tax Bill 2006.

All of the metropolitan trains and most of the metropolitan buses in Adelaide (capital of South Australia) operate on a B5 blend. The South Australian Government has stated that it will soon move to B20 or possibly higher blends.

Several councils (local Governments) across Australia are using B20 (including Townsville City Council, Adelaide City Council, Sydney City Council and Newcastle City Council).

Brazil

Brazil opened a commercial biodiesel refinery in March 2005. It is capable of producing 12,000 m^3 (3.2 million US gallons) per year of biodiesel fuel. Feedstocks can be a variety of sunflower oil, soybean oil, or castor bean oil. The finished product will be a blend of diesel fuel with 2% biodiesel and, after 2013, 5% biodiesel, both usable in unmodified diesel engines. As of 2005, there were 3 refineries and 7 that are planned to open. These three factories were capable of producing 45.6 million of liters per year.

Petrobras (the Brazilian national petroleum company) launched an innovative system, making biodiesel (called H-Bio) from the petroleum refinery. In Brazil, castor bean is the best option to make biodiesel, because it's easier to plant and costs less than soybean, sunflower or other seeds.

Belgium

In Belgium, there are refineries in Ertvelde (belonging to the company Oleon) and at Feluy.

Cambodia

Phnom Penh Biodiesel Cambodia started producing biodiesel in 2007 using a 100-liter biodiesel batch processor imported from Thailand. Biodiesel Cambodia focused on biodiesel production from Jatropha Curcus plants. They originally worked on grass roots projects, encouraging farmers to plant Jatropha. In 2008 Biodiesel Cambodia started focusing on large scale plantations and attempted to partner with large scale land owners and investors. When oil prices crashed suddenly investors pulled out and Biodiesel Cambodia sold their production equipment and ceased operations in late 2008.

Sihanoukville in 2007 a non-profit named Planet Biodiesel Outreach Cambodia started a small scale biodiesel production plant. The biodiesel was used to operate their school bus and to provide fuel for their travel company. Their idea was to operate an environmentally friendly tour company and to use the proceeds to run a school for impoverished Cambodian children and provide food, clothing, education, school supplies and transportation free of charge. They ran their school bus on 100% biodiesel that was produced from waste vegetable oil. Planet Biodiesel Outreach Cambodia shut down in 2008 due to lack of donor funding and the inability to raise enough revenue from their tour company.

Siem Reap - In 2008 Angkor Hospital for Children decided they wanted to power their new visitor center with 100% biodiesel in an effort to be environmentally responsible. A new local NGO was formed called Naga Biofuels to produce the biodiesel from used cooking oil. Naga Biofuels worked with Angkor Hospital and also partnered with several other non-profits in Cambodia to produce biodiesel and a glycerin based soap from used cooking oil. In 2011 Naga Biofuels started working with Angkor Golf Resort and several other businesses and expanded their operations.

After 7 years of being the only biodiesel producer in Cambodia, Naga Biofuels changed their name to Naga Earth and started focusing on additional recycling projects in the Kingdom of Cambodia. As of July 2016, they currently are partnering with 25 different businesses and NGOs that are using their biodiesel.

Canada

The Government of Canada exempted biodiesel from the federal excise tax on diesel in the March 2003 budget. However Government of Canada re-introduced federal excise tax on Bio-Diesel in the March 2013 budget and takes effect on April 1, 2014.

- Quebec - Rothsay of Ville Ste Catherine, Quebec, produces 35,000 m^3 of biodiesel per year. The shuttle bus connecting students between the two campuses of Concordia University are run solely on Biodiesel. Targray, one of the most important B99 and B100 biodiesel suppliers in North America, is headquartered in Kirkland, Quebec, an on-island suburb of Montreal.
- Nova Scotia - The Provincial Government of Nova Scotia uses biodiesel in some public buildings for heating as well as (in more isolated cases) for public transportation. Halifax Regional Municipality has converted its bus fleet to biodiesel, with a future demand of 7,500 m^3 of B20 (20% biodiesel fuel mixture) to B50—reducing biodiesel content in low temperatures to avoid gelling issues—and 3,000 m^3 split between B20 and B100 for building heat. The municipality forecasts a greenhouse gas reduction of over 9,000 tonnes CO_2 equivalents (4,250 tonnes from fleet use and 5,000 tonnes from building heating) if fully implemented. Private sector uptake is slower—but not unheard of—possibly

due to a lack of price differential with petroleum fuel and a lack of federal and provincial tax rebating. Ocean Nutrition Canada produces 6 million US gallons (23,000 m^3) of fatty acid ethyl esters annually as a byproduct of its Omega-3 fatty acid processing.

- Ontario - Great Lakes Biodiesel the Largest Producer of Biodiesel in Canada. Annual Production: 170 million liters per year. Completed and started production in October 2013.

Methes Energies Canada Inc. of Mississauga is processing out of two facilities (Sombra and Mississauga) with a production capacity of 55 million liters of biodiesel per year. Biox Corporation of Oakville is building a biodiesel processing plant in the Hamilton harbor industrial lands, due for completion in the first half of 2006. There are also a few retail filling stations selling biodiesel to motorists in Toronto and Unionville. SIL-TRI Biofuels in Hamilton also delivers B1-B100 blends in bulk for commercial users.

- Manitoba A rush of building of biodiesel plants in 2005 and 2006 started in June 2005 with Bifrost Bio-Diesel in Arborg. In addition, biodiesel is made by individuals and farmers for personal use.
- British Columbia - the cooperative association proves a successful structure for micro-economy-of-scale biodiesel production reaching the end-user. Vancouver Biodiesel Co-op, Nelson Biodiesel Co-op, WISE Energy and Island Biodiesel Co-op are notable examples.

The Canadian government has stated a goal of producing 500 million liters of biodiesel by 2010.*"Welcome to Canada Clean Fuels"*.

China

At least two publicly traded companies, China Clean Energy, Inc. and Gushan, manufacture and sell significant amounts of biodiesel in China.

Costa Rica

Costa Rica is a large producer of crude palm oil and this has spurred interest in biodiesel. Currently several small biodiesel production projects are starting in the country. There are also biodiesel reactor manufacturers in Costa Rica which provide equipment to the Central American and Caribbean region.

Czech Republic

Czech production of biodiesel was already above 60,000 m^3 per year by the early 1990s and is now even larger. Many of the plants are very large, including one in Olomouc which produces almost 40,000 m^3 per year. From the summer of 2004, Czech producers of biodiesel for blend receive a subsidy of roughly CEK 9.50/kg. All Škoda diesels built since 1996 are warrantied for DIN EN 590 biodiesel use.

Estonia

Biodiesel is available at some fuel stations.

European Union

According to EU Strategy for Biofuels at year 2010 target is 5.75% market share for biofuels. At year 2020 the target is 10% market share.

European standard DIN EN 14214 and DIN EN 590 describes the physical properties of diesel fuels.

The EU and environmental organization e.g. Greenpeace demand transparent criteria for sustainable biofuel production. Concern includes deforestation for oil palms in Indonesia and deforestation in Brazil.

Consumption of Biodiesel in the European Union (GWh)						
#		2005	2006	2007	2008	2010
1	Germany	18,003	29,447	33,800	28,819	25,993
2	France	4,003	6,855	14,121	23,501	23,532
3	Italy	2,000	1,732	1,580	6,481	15,088
4	Spain	270	629	3,012	6,036	13,803
5	United Kingdom	292	1,533	3,147	8,040	9,616
6	Poland	152	491	291	3,961	9,179
7	Austria	920	3,878	2,206	2,171	4,749
8	Portugal	2	818	1,570	1,545	3,783
9	Belgium	0	10	1,061	1,002	3,223
10	Sweden	97	523	1,195	1,511	2,035
11	Czech Republic	33	213	327	881	2,006
12	Romania	-	32	465	700	1,464
13	Greece	32	540	945	880	1,449
14	Slovakia	58	48	896	617	1,408
15	Hungary	0	4	23	942	1,357
16	Netherlands	0	172	2,559	2,349	1,101

17	Ireland	9	8	201	465	694
18	Finland	0	0	1	133	615
19	Slovenia	110	149	151	259	485
20	Luxembourg	7	6	484	482	466
21	Lithuania	87	162	489	532	404
22	Latvia	29	17	20	22	217
23	Bulgaria	n.a.	96	23	342	185
24	Cyprus	0	0	9	165	174
25	Denmark	0	0	0	0	8
-	Estonia	0	7	6	32	0
-	Malta	8	10	21	11	0
	European Union	**26,110**	**47,380**	**68,602**	**91,880**	**123,035**
1 toe = 11.63 MWh, n.a. =not available						

Finland

Neste Oil is producer of NExBTL renewable diesel oil. NExBTL is first renewable fuel suitable for all diesel engines in the world. Neste Oil claims EN-590 outperforms both regular diesel as well as other biodiesels on the market.

France

The French government have ambitious objectives of incorporation of biofuels in fuels: 5,75% (energy value) in 2008, 7% in 2010 and 10% by 2015.

Germany

According to the *Union zur Förderung von Öl- und Proteinpflanzen* UFOP (Union to promote oil- and protein plants), in 2006 the sale of biodiesel through German gas stations rose to 2,000,000 m^3, although it is only available at privately owned filling stations and generally not available at outlets operated by major petroleum companies, such as Shell and Esso/Exxon (the petroleum companies see biodiesel as competition to their core petroleum business). In 2004, 45% of all biodiesel sales went directly to large end users, such as trucking companies.

In Germany biodiesel is, for the most part, produced from rapeseed. Sales in Germany stood at two billion litres (about 600 million US gallons (2,300,000 m^3)) in 2006. This amount was sufficient to meet the average yearly consumption of well over 2,000,000 automobiles. Diesel engines have become increasingly popular in Germany and almost half of all newly manufactured cars are diesel powered. This is in part due to the greater efficiency of diesel engines, the desire by consumers to use environmentally friendlier technologies and lower taxes on diesel fuel that make it cheaper than gasoline.

Verbio is a biodiesel and bioethanol producer in Germany. They currently produce 400,000 tons of biodiesel from rape seed oil, soybean oil and fatty acids, selling the product to markets in the E.U.

With 1,900 sales points, equal to one in every ten public gas stations, biodiesel is the first alternative fuel to be available nationwide. The industry is expecting a surge in demand since the authorisation at the beginning of 2004, through European Union legislation, of a maximum 5% biodiesel addition to conventional diesel fuel. In Germany biodiesel is sold at a lower price than diesel from petroleum.

Hong Kong

Dynamic Progress International is the largest biodiesel producer in Hong Kong, the feedstock for the biodiesel it sells locally, being used cooking oil also sourced locally. In November 2011, Dynamic Progress International was chosen in an open tendering procedure, to supply over 3.5 million litres of B5 diesel to various Hong Kong Government departments on a 16-month contract. Another biodiesel plant is currently under construction in Hong Kong on the Tseung Kwan O Industrial Estate by the ASB Biodiesel (Hong Kong) Limited. Upon completion, ASB Biodiesel will become the biggest biodiesel plant in Hong Kong.

India

Bio Diesel Powered Fast Attack Craft Of Indian Navy

Biodiesel is now being produced locally in India for use in stationary engines and large or slow engines like those in trains, trucks and tractors. Efforts are also on to use ethanol as a substitute for petrol. Biodiesel is rapidly replacing both kerosene (which was used illegally and inefficiently) and diesel as a more efficient, cheap, and clean alternative for large engines. Today plans are being chalked out to cultivate non-edible seed oil trees such as Pongamia on barren land to use their oil for biodiesel production. The Southern Railways have been very supportive of the program and various state governments recommend planting these plants in unused areas through aid from the government sectors. Biodiesel-blends are being used to run state transport corporation buses in Karnataka. The University of Agriculture Sciences at Bangalore has identified many elite lines of *Jatropha curcas* and *Millettia pinnata* (Pongamia tree). Castor is already a well established crop in India with several very high yielding varieties in the market. Large-scale activities have been initiated quite recently. For example, the government of Karnataka has distributed several million saplings of Pongamia to farmers for planting along borders of farmland and in waste lands. Large-scale plantations have been initiated in North-East India and Jharkhand by D1 Williamson Magor Bio Fuel Limited, a joint venture between D1 Oils of U.K. and Williamson Magor of India. The hilly areas of the North-East are ideal for growing hardy, low-maintenance plants.

IndianOil Corporation has tied up with Indian Railways to introduce biodiesel crops over 1 million square kilometers. Also, Jharkhand and Madhya Pradesh have tied up with IndianOil to cultivate large tracts of land with jatropha, the former crop of choice for Indian biodiesel plans. Jatropha is now being replaced by Pongamia and Castor due to its very (comparative) high cost of cultivation. In order to organize the industry, the BioDiesel Society of India has been formed to encourage energy plantations for increasing feedstock supplies.

Indonesia

Since September 2005, the Eterindo Group has been producing biodiesel using palm oil. Currently the production capacity of the Eterindo Group has reached 120,000 tonnes of Biodiesel annually. Meeting the standard requirements of ASTM D-6751 and EN 14214, in 2006 the group begin to export its biodiesel to United States, Germany and Japan. It is now exploring another export destinations, i.e.: Asia Pacific countries, etc.

Israel

Biodiesel is not yet sold on the market, things start to change and biodiesel is being produced in two small-scale experiments. The amounts produced in these experiments are up to 10,000 liters a month. The lack of production of biodiesel in Israel is in contrary with the Research and Development abilities of the country, for Israel is a center of development for agriculture technologies. The Israel North Recycle Group (INRG) is forecasting much progress in the next year, including consumption agreements with municipal bodies, as part of the wider view of the municipalities on the subject.

Italy

Italy has the capacity to produce over 2,711 million litres of biodiesel each year, but due to competition from imports, production has decreased to about 500,000 tonnes in 2012. Consumption was 1,681 million litres in 2011, and was expected to be 1,573 million litres in 2012.

Malaysia

The biodiesel industry has been identified as one of the 12 National Key Economic Activities (NKEA) under the Tenth Malaysia Plan.

Projects requiring Malaysian and Indonesian palm oil as feedstocks have been criticized by some environmental advocates. Friends of the Earth has published a report asserting that clearance of forests for oil-palm plantations is threatening some of the last habitat of the orangutan. Also, in a column for The Guardian, writer George Monbiot claimed that land clearance by cutting and burning large forest trees frees large amounts of carbon dioxide that is never reabsorbed by the smaller oil palms. If true, then biodiesel production from plantation-grown palm oil may be a net source of atmospheric carbon dioxide. How these issues are resolved may determine whether Malaysia eventually becomes a major producer of biodiesel.

The palm oil industry has recognized this concerned and in conjunction with the WWF has formed the Roundtable on Sustainable Palm Oil (RSPO) which endeavours to ensure development of palm oil in a sustainable way. In 2014, after many delays Malaysia began the introduction in the sale of B5 biodiesel in most petrol stations in the country. This would be eventually replaced by B7 and finally B10 biodiesel blends in late 2015.

Lithuania

There are two biodiesel producing plants in Lithuania. One in close to Mažeikiai (Samogitia) and other in Klaipėda free economic zone. Most of biodiesel produced in Lithuania is consumed at the local market and a fraction exported.

Norway

Biodiesel has been launched across all of Norway. Starting from 2008, the B5 diesel has been the standard fuel at almost all stations, and the transformation from normal diesel will be completed within 2008

Pakistan

In Pakistan Biodiesel has been launched by Pakistan State Oil in the Sindh province. Although details about the Biodeisel are still cloudy as there was almost no media campaign, the mix used is E10.

Currently ethanol is being derived from Sugar Mills, that produce it as a by product and in many cases, use it to power their own mills.

Papua New Guinea

Biodiesel is produced from Copra oil (oil extracted from coconuts) in a processing plant at Buka on the island of Bougainville, vehicles which run on this fuel have a sticker on the doors which says "powered by Coconuts". This fuel is cheaper and more readily available than imported PetroDiesel. The oil-company StatoilHydro has recently started to distribute biodiesel too.

Spain

It is possible to buy biodiesel, mixed with diesel fuel, in more than 480 petrol stations around the country.

Singapore

Two biodicsel plants will be built on Jurong Island, Singapore's petrochemicals hub. The first plant, by Peter Cremer (S) GMBH, will have a capacity of 200,000 tons/year and it is expected to be ready by early 2007, while the second is a joint venture between Wilmar Holdings and Archer Daniels Midland Company, to be operational by end 2006 with an initial capacity of 150,000 tons/year.

Natural Fuel Pte Ltd, has mechanically completed a 600,000 tons/year biodiesel plant in early 2008 - making it one of the world's largest biodiesel located in a single site.

Singapore was selected for the companies' first biodiesel plant in Asia because of its excellent connectivity. There is easy access to abundant palm oil feedstock from the neighboring countries of Malaysia and Indonesia. Also, Singapore has terminal facilities which allow the biodiesel to be shipped to markets around the world.

Neste Oil is building NExBTL renewable diesel oil plant, production 800,000 tons/ year. According plan it will be ready 2010.

Taiwan

In 2004, several companies started making biodiesel, and has produced more than 5,000 kilotons in a year since then. In 2006, the Bureau of Energy launched the first biodiesel buses on Earth Day.

Thailand

Thailand was the first country to launch biodiesel as a national program on July 10, 2001. It was reported that the work was initiated by the Royal Chitralada Project, a royal -sponsored project to help rural farmers . International co-operation among ASEAN country was also starting by the Renewable Energy Institute of Thailand (Dr. Samai Jai-In) and Asia-Pacific Roundtable for Sustainable consumption and Production (Dr. Olivia Castillo,). The primary aims of the project in Thailand are:

- an alternative output for excess agricultural produce
- substituting diesel imports

In 2007, several biodiesel plants are operating in Thailand using the excess palm oil / palm stearin and in some cases, waste vegetable oil as raw materials. The production capacity is about 1 million litre/day and should reach 2 million litre by early 2008. About 400 petrol stations are now distributing B5 (5% biodiesel with 95% diesel) in Chiangmai and Bangkok. The national biodiesel standard has been developed based on the European standard. The target of the Government is to mandate B2 by 2 April 2008 and to increase to B5 by 2011 which will require almost 4 Million litres/day of biodiesel .

The raw material will most likely come from palm oil, coconut oil, Jatropha Curcas Linn, and tallow. Several pilot plants are now operating such as the Royal Chitralada Projects , Rajabiodiesel in Surattani , Department of Alternative Energy Development and Efficiency , Royal Naval Dockyard, MTEC , and Tistr [www.tistr.or.th].

United Kingdom

Biodiesel is sold by a small but growing number of filling stations in B5 and B100 blend, including a significant fraction of supermarket filling stations. Some farmers have also been using small plants to create their own biodiesel for farm machinery since the 1990s. Several Co-ops and small-scale production facilities have recently begun production, typically selling fuel several pence per litre less than petrodiesel. The first large-scale plant, capable of producing 50 million litres (13 million US gallons) a year, opened in Scotland in 2005, soon followed by a large plant co-owned by Tesco and Greenergy (Tesco sell 5% biodiesel at many of their petrol stations. The Fuel conforms to standard EN590 which allows up to 5% biodiesel inclusion). Biodiesel is treated like any other vehicle fuel in the UK and the paperwork required to register as a producer is a major limiting factor to growth in the market. Although since July 2007 home users may produce 2500 litres per year for personal use without registering or paying duty.

United States

Biodiesel is commercially available in most oilseed-producing states in the United States. As of 2005, it is somewhat more expensive than fossil diesel, though it is still commonly produced in relatively small quantities (in comparison to petroleum products and ethanol). Many farmers who raise oilseeds use a biodiesel blend in tractors and equipment as a matter of policy, to foster production of biodiesel and raise public awareness. It is sometimes easier to find biodiesel in rural areas than in cities. Similarly, some agribusinesses and others with ties to oilseed farming use biodiesel for public relations reasons. As of 2003 some tax credits were available in the U.S. for using biodiesel. In 2004 almost 30 million US gallons (110,000 m^3) of commercially produced biodiesel were sold in the U.S., up from less than 0.1 million US gallons (380 m^3) in 1998. Projections for 2005 were 75 million US gallons (280,000 m^3) produced from 45 factories and 150 million US gallons (570,000 m^3). Due to increasing pollution control requirements and tax relief, the U.S. market is expected to grow to 1 billion US gallons (3,800,000 m^3) or 2 billion US gallons (7,600,000 m^3) by 2010.

Uruguay

Uruguayan law 18.195 stipulates a minimum of 2% of biodiesel in diesel since January 2009 and 5% from January 2012. As of March 2009, state fuel monopoly ANCAP pretends to start blending biodiesel for automotive use in late May or early June 2009.

Ester

In chemistry, esters are chemical compounds derived from an acid (organic or inorganic) in which at least one –OH (hydroxyl) group is replaced by an –O–alkyl (alkoxy) group. Usually, esters are derived from a carboxylic acid and an alcohol. Glycerides, which are fatty acid esters of glycerol, are important esters in biology, being one of the main classes of lipids, and making up the bulk of animal fats and vegetable oils. Esters with low molecular weight are commonly used as fragrances and found in essential oils and pheromones. Phosphoesters form the backbone of DNA molecules. Nitrate esters, such as nitroglycerin, are known for their explosive properties, while polyesters are important plastics, with monomers linked by ester moieties.

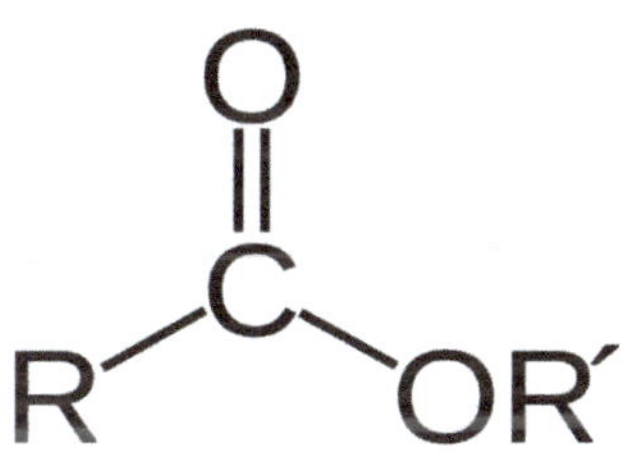

A carboxylate ester. R and R′ denote any alkyl or aryl group. R can also be a hydrogen atom

Nomenclature

Etymology

The word 'ester' was coined in 1848 by German chemist Leopold Gmelin, probably as a contraction of the German *Essigäther*, "acetic ether".

IUPAC Nomenclature

Ester names are derived from the parent alcohol and the parent acid, where the latter may be organic or inorganic. Esters derived from the simplest carboxylic acids are commonly named according to the more traditional, so-called "trivial names" e.g. as formate, acetate, propionate, and butyrate, as opposed to the IUPAC nomenclature methanoate, ethanoate, propanoate and butanoate. Esters derived from more complex carboxylic acids are, on the other hand, more frequently named using the systematic IUPAC name, based on the name for the acid followed by the suffix *-oate*. For example, the ester hexyl octanoate, also known under the trivial name hexyl caprylate, has the formula $CH_3(CH_2)_6CO_2(CH_2)_5CH_3$.

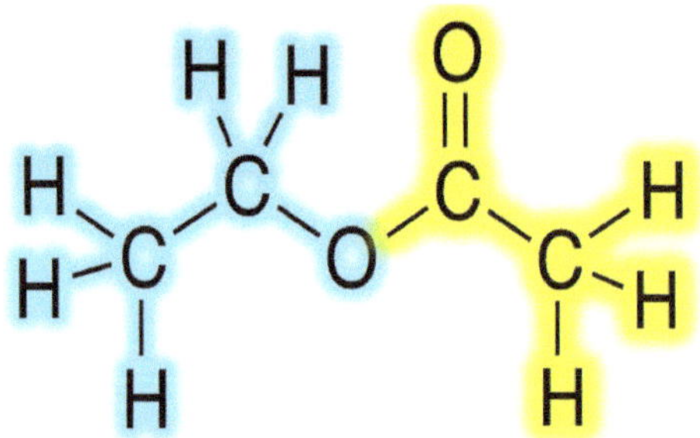

Ethyl acetate derived from an alcohol (blue) and an acyl group (yellow) derived from a carboxylic acid.

The chemical formulas of organic esters usually take the form RCO_2R', where R and R′ are the hydrocarbon parts of the carboxylic acid and the alcohol, respectively. For example, butyl acetate (systematically butyl ethanoate), derived from butanol and acetic acid (systematically ethanoic acid) would be written $CH_3CO_2C_4H_9$. Alternative presentations are common including BuOAc and $CH_3COOC_4H_9$.

Cyclic esters are called lactones, regardless of whether they are derived from an organic or an inorganic acid. One example of a (organic) lactone is γ-valerolactone.

Orthoesters

An uncommon class of organic esters are the orthoesters, which have the formula $RC(OR')_3$. Triethylorthoformate ($HC(OC_2H_5)_3$) is derived, in terms of its name (but not its synthesis) from orthoformic acid ($HC(OH)_3$) and ethanol.

Inorganic Esters

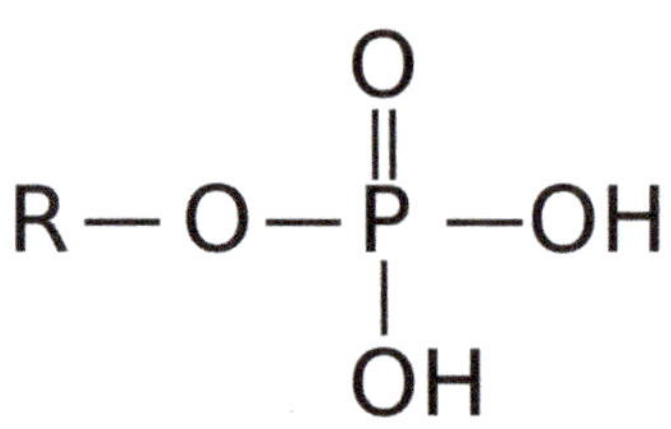

A phosphoric acid ester

Esters can also be derived from an inorganic acid and an alcohol. Thus, the nomenclature extends to inorganic oxo acids and their corresponding esters: phosphoric acid and phosphate esters/organophosphates, sulfuric acid and sulfate esters/organosulfates, nitric acid and nitrate, and boric acid and borates. For example, triphenyl phosphate is the ester derived from phosphoric acid and phenol. Organic carbonates are derived from carbonic acid; for example, ethylene carbonate is derived from carbonic acid and ethylene glycol.

So far an alcohol and inorganic acid are linked via oxygen atoms. The definition of inorganic acid ester that feature inorganic chemical elements links between alcohols and the inorganic acid – the phosphorus atom linking to three alkoxy functional groups in organophosphate – can be extended to the same elements in various combinations of covalent bonds between carbons and the central inorganic atom and carbon–oxygen bonds to central inorganic atoms. For example, phosphorus features three carbon–oxygen–phosphorus bondings and one phosphorus–oxygen double bond in organophosphates,

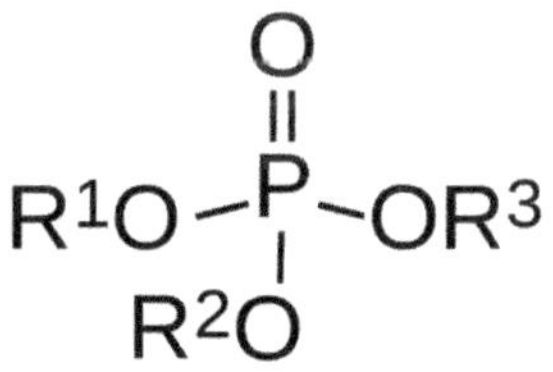

structure of a generic organophosphate

three carbon–oxygen–phosphorus bondings and no phosphorus–oxygen double bonds in phosphite esters or organophosphites,

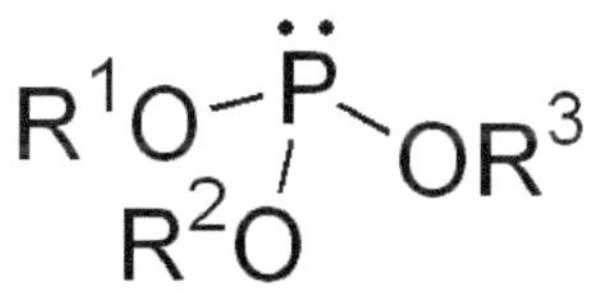

structure of a generic phosphite ester showing the lone pairs on the P

two carbon–oxygen–phosphorus bondings, no phosphorus–oxygen double bonds but one phosphorus–carbon bond in phosphonites,

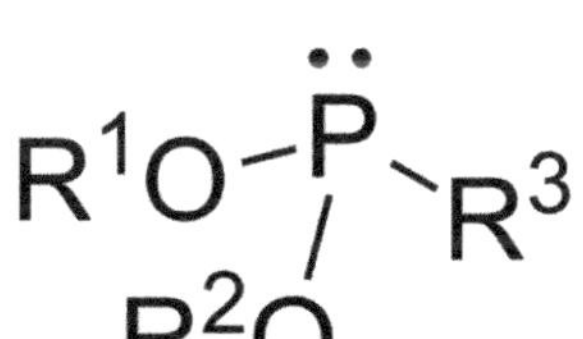

structure of a generic phosphonite – ester of phosphonous acid

one carbon–oxygen–phosphorus bondings, no phosphorus–oxygen double bonds but two phosphorus–carbon bonds in phosphinites.

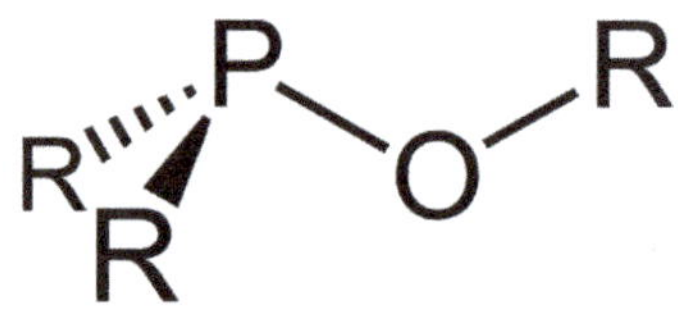

structure of a generic phosphinite.
In corollary, boron features borinic esters (n = 2), boronic esters (n = 1), and borates (n = 0).

As oxygen is a group 16 chemical element, sulfur atoms can replace some oxygen atoms in carbon–oxygen–central inorganic atom covalent bonds of an ester. As a result, thiosulfinates' and thiosulfonates, with a central inorganic sulfur atom, demonstrate clearly the assortment of sulfur esters, that also includes sulfates, sulfites, sulfonates, sulfinates, sulfenates esters.

Structure and Bonding

Esters contain a carbonyl center, which gives rise to 120 ° C–C–O and O–C–O angles. Unlike amides, esters are structurally flexible functional groups because rotation about the C–O–C bonds has a low barrier. Their flexibility and low polarity is manifested in their physical properties; they tend to be less rigid (lower melting point) and more volatile (lower boiling point) than the corresponding amides. The pK_a of the alpha-hydrogens on esters is around 25.

Many esters have the potential for conformational isomerism, but they tend to adopt an *s*-cis (or Z) conformation rather than the *s*-trans (or E) alternative, due to a combination of hyperconjugation and dipole minimization effects. The preference for the Z conformation is influenced by the nature of the substituents and solvent, if present. Lactones with small rings are restricted to the *s*-trans (i.e. E) conformation due to their cyclic structure.

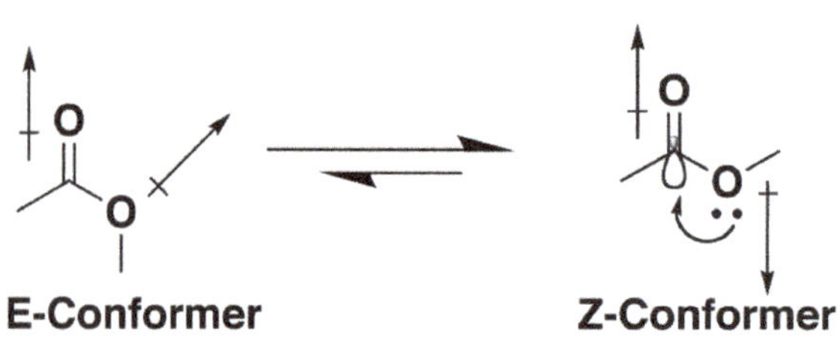

Physical Properties and Characterization

Esters are more polar than ethers but less polar than alcohols. They participate in hydrogen bonds as hydrogen-bond acceptors, but cannot act as hydrogen-bond donors, unlike their parent alcohols. This ability to participate in hydrogen bonding confers some water-solubility. Because of their lack of hydrogen-bond-donating ability, esters do not self-associate. Consequently, esters are more volatile than carboxylic acids of similar molecular weight.

Characterization and Analysis

Esters are generally identified by gas chromatography, taking advantage of their volatility. IR spectra for esters feature an intense sharp band in the range 1730–1750 cm^{-1} assigned to $\nu C{=}O$. This peak changes depending on the functional groups attached to the carbonyl. For example, a benzene ring or double bond in conjugation with the carbonyl will bring the wavenumber down about 30 cm^{-1}.

Applications and Occurrence

Esters are widespread in nature and are widely used in industry. In nature, fats are in general triesters derived from glycerol and fatty acids. Esters are responsible for the aroma of many fruits, including apples, durians, pears, bananas, pineapples, and strawberries. Several billion kilograms of polyesters are produced industrially annually, important products being polyethylene terephthalate, acrylate esters, and cellulose acetate.

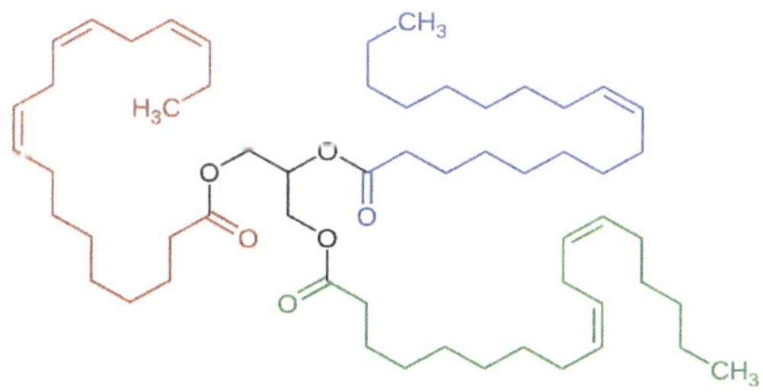

Representative triglyceride found in a linseed oil, a triester (triglyceride) derived of linoleic acid, alpha-linolenic acid, and oleic acid.

Preparation

Esterification is the general name for a chemical reaction in which two reactants (typically an alcohol and an acid) form an ester as the reaction product. Esters are common in organic chemistry and biological materials, and often have a characteristic pleasant, fruity odor. This leads to their extensive use in the fragrance and flavor industry. Ester bonds are also found in many polymers.

Esterification of Carboxylic Acids

The classic synthesis is the Fischer esterification, which involves treating a carboxylic acid with an alcohol in the presence of a dehydrating agent:

$$RCO_2H + R'OH \rightleftharpoons RCO_2R' + H_2O$$

The equilibrium constant for such reactions is about 5 for typical esters, e.g., ethyl acetate. The reaction is slow in the absence of a catalyst. Sulfuric acid is a typical catalyst for this reaction. Many other acids are also used such as polymeric sulfonic acids. Since esterification is highly reversible, the yield of the ester can be improved using Le Chatelier's principle:

- Using the alcohol in large excess (i.e., as a solvent).
- Using a dehydrating agent: sulfuric acid not only catalyzes the reaction but sequesters water (a reaction product). Other drying agents such as molecular sieves are also effective.
- Removal of water by physical means such as distillation as a low-boiling azeotropes with toluene, in conjunction with a Dean-Stark apparatus.

Reagents are known that drive the dehydration of mixtures of alcohols and carboxylic acids. One example is the Steglich esterification, which is a method of forming esters under mild conditions. The method is popular in peptide synthesis, where the substrates are sensitive to harsh conditions like high heat. DCC (dicyclohexylcarbodiimide) is used to activate the carboxylic acid to further reaction. DMAP (4-dimethylaminopyridine) is used as an acyl-transfer catalyst.

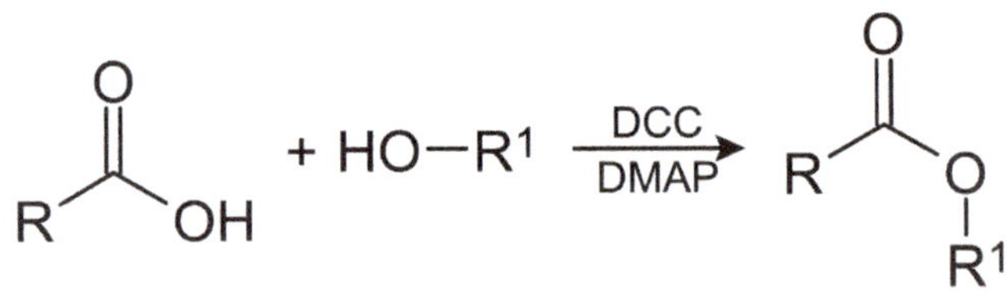

Another method for the dehydration of mixtures of alcohols and carboxylic acids is the Mitsunobu reaction:

$$RCO_2H + R'OH + P(C_6H_5)_3 + R_2N_2 \rightarrow RCO_2R' + OP(C_6H_5)_3 + R_2N_2H_2$$

Carboxylic acids can be esterified using diazomethane:

$$RCO_2H + CH_2N_2 \rightarrow RCO_2CH_3 + N_2$$

Using this diazomethane, mixtures of carboxylic acids can be converted to their methyl esters in near quantitative yields, e.g., for analysis by gas chromatography. The method is useful in specialized organic synthetic operations but is considered too hazardous and expensive for large scale applications.

Alcoholysis of Acyl Chlorides and Acid Anhydrides

Alcohols react with acyl chlorides and acid anhydrides to give esters:

$$RCOCl + R'OH \rightarrow RCO_2R' + HCl$$

$$(RCO)_2O + R'OH \rightarrow RCO_2R' + RCO_2H$$

The reactions are irreversible simplifying work-up. Since acyl chlorides and acid anhydrides also react with water, anhydrous conditions are preferred. The analogous acylations of amines to give amides are less sensitive because amines are stronger nucleophiles and react more rapidly than does water. This method is employed only for laboratory-scale procedures, as it is expensive.

Alkylation of Carboxylate Salts

Although not widely employed for esterifications, salts of carboxylate anions can be alkylating agent with alkyl halides to give esters. In the case that an alkyl chloride is used, an iodide salt can catalyze the reaction (Finkelstein reaction). The carboxylate salt is often generated *in situ*. In difficult cases, the silver carboxylate may be used, since the silver ion coordinates to the halide aiding its departure and improving the reaction rate. This reaction can suffer from anion availability problems and, therefore, can benefit from the addition of phase transfer catalysts or highly polar aprotic solvents such as DMF.

Transesterification

Transesterification, which involves changing one ester into another one, is widely practiced:

$$RCO_2R' + CH_3OH \rightarrow RCO_2CH_3 + R'OH$$

Like the hydrolysation, transesterification is catalysed by acids and bases. The reaction is widely used for degrading triglycerides, e.g. in the production of fatty acid esters and alcohols. Poly(ethylene terephthalate) is produced by the transesterification of dimethyl terephthalate and ethylene glycol:

$$(C_6H_4)(CO_2CH_3)_2 + 2\ C_2H_4(OH)_2 \rightarrow \tfrac{1}{n}\ \{(C_6H_4)(CO_2)_2(C_2H_4)\}_n + 2\ CH_3OH$$

Carbonylation

Alkenes undergo "hydroesterification" in the presence of metal carbonyl catalysts. Esters of propionic acid are produced commercially by this method:

$$C_2H_4 + ROH + CO \rightarrow C_2H_5CO_2R$$

The carbonylation of methanol yields methyl formate, which is the main commercial source of formic acid. The reaction is catalyzed by sodium methoxide:

$$CH_3OH + CO \rightarrow CH_3O_2CH$$

Addition of Carboxylic Acids to Alkenes

In the presence of palladium-based catalysts, ethylene, acetic acid, and oxygen react to give vinyl acetate:

$$C_2H_4 + CH_3CO_2H + \tfrac{1}{2}\ O_2 \rightarrow C_2H_3O_2CCH_3 + H_2O$$

Direct routes to this same ester are not possible because vinyl alcohol is unstable.

Other Methods

- Favorskii rearrangement of α-haloketones in presence of base
- Baeyer–Villiger oxidation of ketones with peroxides
- Pinner reaction of nitriles with an alcohol
- Nucleophilic abstraction of a metal–acyl complex
- Hydrolysis of orthoesters in aqueous acid
- Cellulolysis via esterification
- Ozonolysis of alkenes using a work up in the presence of hydrochloric acid and various alcohols.
- Anodic oxidation of methyl ketones leading to methyl esters.

Reactions

Esters react with nucleophiles at the carbonyl carbon. The carbonyl is weakly electrophilic but is attacked by strong nucleophiles (amines, alkoxides, hydride sources, organolithium compounds, etc.). The C–H bonds adjacent to the carbonyl are weakly acidic but undergo deprotonation with strong bases. This process is the one that usually initiates condensation reactions. The carbonyl oxygen is weakly basic (less so than in amides) but forms adducts.

Addition of Nucleophiles at Carbonyl

Esterification is a reversible reaction. Esters undergo hydrolysis under acid and basic conditions. Under acidic conditions, the reaction is the reverse reaction of the Fischer esterification. Under basic conditions, hydroxide acts as a nucleophile, while an alkoxide is the leaving group. This reaction, saponification, is the basis of soap making.

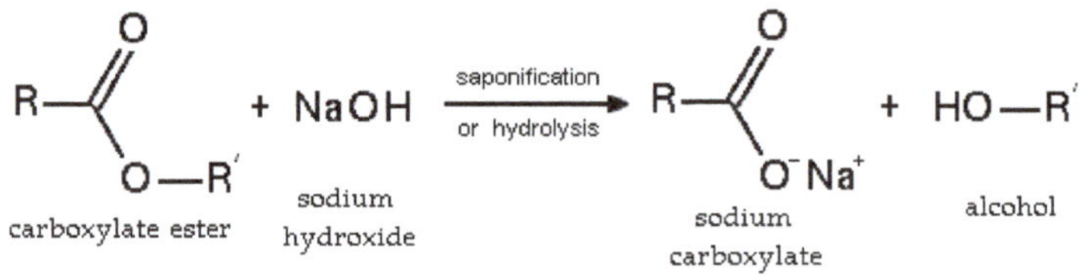

The alkoxide group may also be displaced by stronger nucleophiles such as ammonia or primary or secondary amines to give amides: (ammonolysis reaction)

$$RCO_2R' + NH_2R'' \rightarrow RCONHR'' + R'OH$$

This reaction is not usually reversible. Hydrazines and hydroxylamine can be used in place of amines. Esters can be converted to isocyanates through intermediate hydroxamic acids in the Lossen rearrangement.

Sources of carbon nucleophiles, e.g., Grignard reagents and organolithium compounds, add readily to the carbonyl.

Reduction

Compared to ketones and aldehydes, esters are relatively resistant to reduction. The introduction of catalytic hydrogenation in the early part of the 20th century was a breakthrough; esters of fatty acids are hydrogenated to fatty alcohols.

$$RCO_2R' + 2\ H_2 \rightarrow RCH_2OH + R'OH$$

A typical catalyst is copper chromite. Prior to the development of catalytic hydrogenation, esters were reduced on a large scale using the Bouveault–Blanc reduction. This method, which is largely obsolete, uses sodium in the presence of proton sources.

Especially for fine chemical syntheses, lithium aluminium hydride is used to reduce esters to two primary alcohols. The related reagent sodium borohydride is slow in this reaction. DIBAH reduces esters to aldehydes.

Direct reduction to give the corresponding ether is difficult as the intermediate hemiacetal tends to decompose to give an alcohol and an aldehyde (which is rapidly reduced to give a second alcohol). The reaction can be achieved using triethylsilane with a variety of Lewis acids.

Claisen Condensation and Related Reactions

As for aldehydes, the hydrogen atoms on the carbon adjacent ("α to") the carboxyl group in esters are sufficiently acidic to undergo deprotonation, which in turn leads to a variety of useful reactions. Deprotonation requires relatively strong bases, such as alkoxides. Deprotonation gives a nucleophilic enolate, which can further react, e.g., the Claisen condensation and its intramolecular equivalent, the Dieckmann condensation. This conversion is exploited in the malonic ester synthesis, wherein the diester of malonic acid reacts with an electrophile (e.g., alkyl halide), and is subsequently decarboxylated. Another variation is the Fráter–Seebach alkylation.

Other Reactions

- Phenyl esters react to hydroxyarylketones in the Fries rearrangement.
- Specific esters are functionalized with an α-hydroxyl group in the Chan rearrangement.
- Esters with β-hydrogen atoms can be converted to alkenes in ester pyrolysis.
- A direct conversion of esters to nitriles.

Protecting Groups

As a class, esters serve as protecting groups for carboxylic acids. Protecting a carboxylic acid is useful in peptide synthesis, to prevent self-reactions of the bifunctional amino acids. Methyl and ethyl esters are commonly available for many amino acids; the *t*-butyl ester tends to be more expensive. However, *t*-butyl esters are particularly useful because, under strongly acidic conditions, the *t*-butyl esters undergo elimination to give the carboxylic acid and isobutylene, simplifying work-up.

List of Ester Odorants

Many esters have distinctive fruit-like odors, and many occur naturally in the essential oils of plants. This has also led to their commonplace use in artificial flavorings and fragrances when those odors aim to be mimicked.

Ester name	Formula	Odor or occurrence
Allyl hexanoate	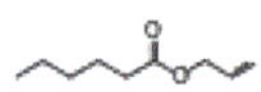	pineapple
Benzyl acetate	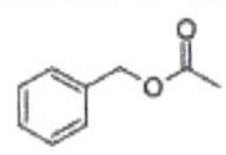	pear, strawberry, jasmine
Bornyl acetate		pine
Butyl acetate	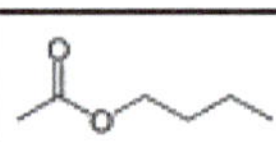	apple, honey
Butyl butyrate	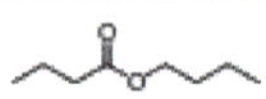	pineapple
Butyl propanoate		pear drops
Ethyl acetate	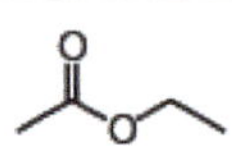	nail polish remover, model paint, model airplane glue
Ethyl benzoate	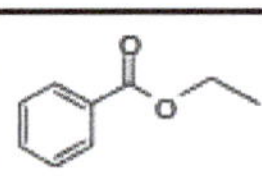	sweet, wintergreen, fruity, medicinal, cherry, grape
Ethyl butyrate	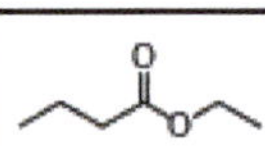	banana, pineapple, strawberry
Ethyl hexanoate	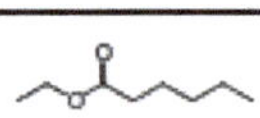	pineapple, waxy-green banana
Ethyl cinnamate	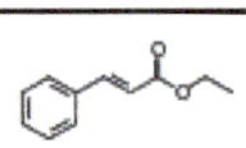	cinnamon
Ethyl formate	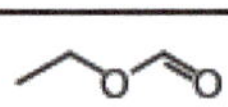	lemon, rum, strawberry
Ethyl heptanoate		apricot, cherry, grape, raspberry

Ethyl isovalerate	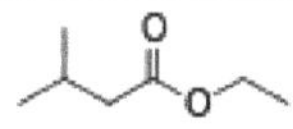	apple
Ethyl lactate	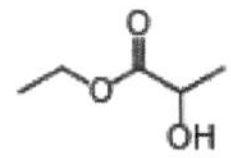	butter, cream
Ethyl nonanoate		grape
Ethyl pentanoate	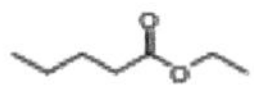	apple
Geranyl acetate	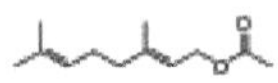	geranium
Geranyl butyrate	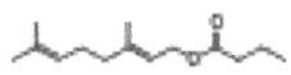	cherry
Geranyl pentanoate		apple
Isobutyl acetate	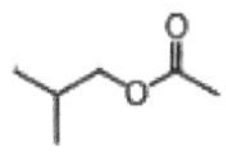	cherry, raspberry, strawberry
Isobutyl formate	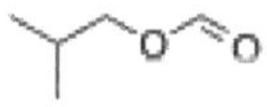	raspberry
Isoamyl acetate	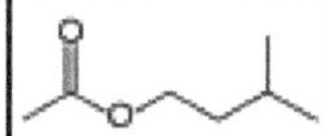	pear, banana (flavoring in Pear drops)
Isopropyl acetate	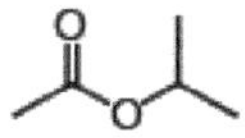	fruity
Linalyl acetate	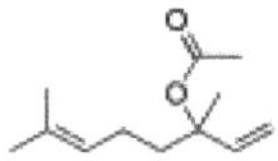	lavender, sage
Linalyl butyrate	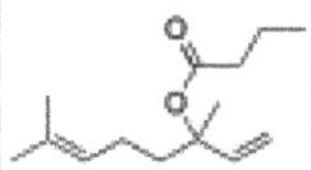	peach
Linalyl formate	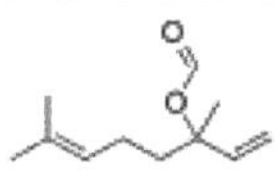	apple, peach
Methyl acetate	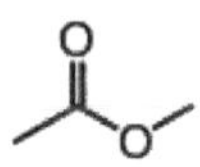	glue
Methyl anthranilate	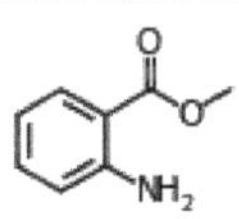	grape, jasmine
Methyl benzoate	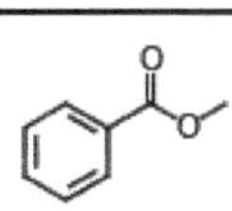	fruity, ylang ylang, feijoa
Methyl butyrate (methyl butanoate)	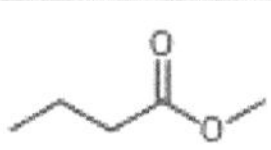	pineapple, apple, strawberry

Methyl cinnamate	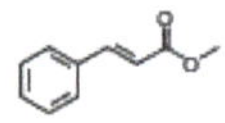	strawberry
Methyl pentanoate (methyl valerate)	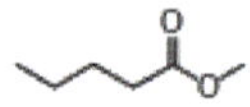	flowery
Methyl phenylacetate	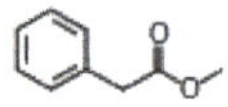	honey
Methyl salicylate (oil of wintergreen)	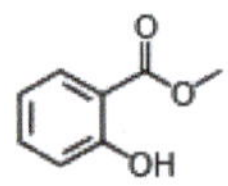 	Modern root beer, wintergreen, Germolene and Ralgex ointments (UK)
Nonyl caprylate	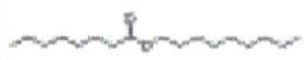	orange
Octyl acetate		fruity-orange
Octyl butyrate		parsnip
Amyl acetate (pentyl acetate)	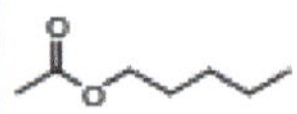	apple, banana
Pentyl butyrate (amyl butyrate)	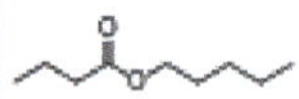	apricot, pear, pineapple
Pentyl hexanoate (amyl caproate)	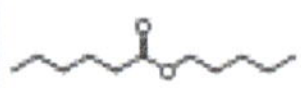	apple, pineapple
Pentyl pentanoate (amyl valerate)	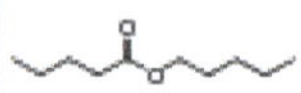	apple
Propyl acetate	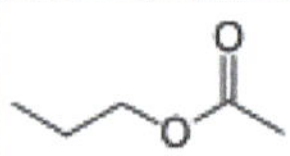	pear
Propyl hexanoate	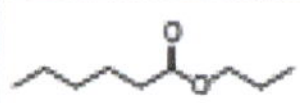	blackberry, pineapple, cheese, wine
Propyl isobutyrate	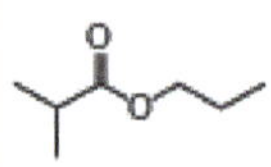	rum
Terpenyl butyrate	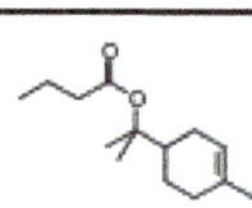	cherry

Alkyl

In organic chemistry, an alkyl substituent is an alkane missing one hydrogen. The term alkyl is intentionally unspecific to include many possible substitutions. An acyclic alkyl has the general formula C_nH_{2n+1}. A cycloalkyl is derived from a cycloalkane by removal of a hydrogen atom from a ring and has the general formula C_nH_{2n-1}. Typically an alkyl is a part of a larger molecule. In structural formula, the symbol R is used to designate a generic (unspecified) alkyl group. The smallest alkyl group is methyl, with the formula CH_3–.

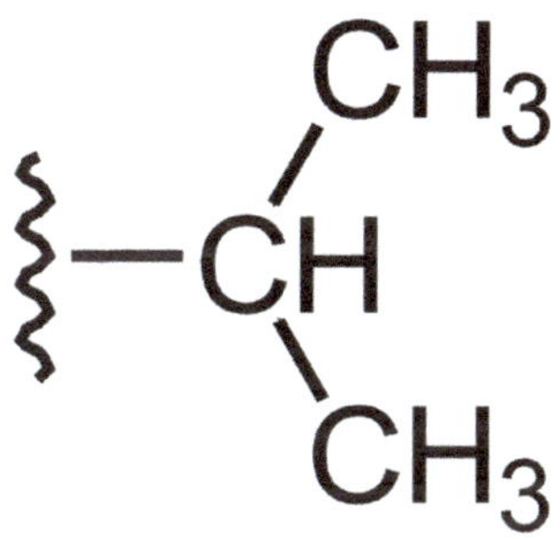

Isopropyl group

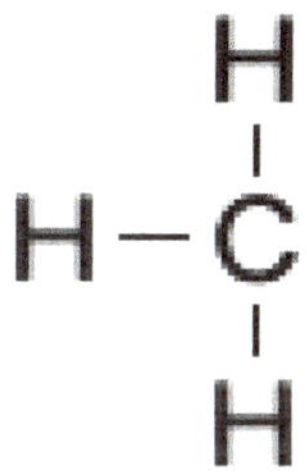

Methyl group

Alkyls in Everyday Life

The word root *alkyl* is encountered in several contexts. Alkylation is an important operation in refineries, for example in the production of high-octane gasoline. Alkylating antineoplastic agents refer to a class of compounds that are used to treat cancer. In such case, the term alkyl is used loosely. For example, nitrogen mustards are well-known alkylating agents, but they are more complex than a mere hydrocarbon. In chemistry, alkyl refers to a group, a substituent, that is attached to other molecular fragments. For example, alkyl lithium reagents have the empirical formula Li(alkyl), where alkyl = methyl, ethyl, etc. A *dialkyl* ether is an ether with two alkyl groups, e.g., diethyl ether ($O(C_2H_5)_2$).

Alkyls in Medicinal Chemistry

In medicinal chemistry, the incorporation of alkyl chains into some chemical compounds increases their lipophilicity. This strategy has been used to increase the antimicrobial activity of flavanones and chalcones.

Alkyl Cations, Anions, and Radicals

Usually alkyl groups are attached to other atoms or groups of atoms. Free alkyls occur as neutral compounds, as anions, or as cations. The cations are called carbocations. The anions are called carbanions. The neutral alkyl free radicals have no special name. Such species are usually encountered only as transient intermediates, but some are quite stable and can be "put into a bottle". Typically alkyl cations are generated using super acids, alkyl anions are observed in the presence of strong bases, and alkyl radicals are generated by a photochemical reaction. Alkyls are commonly observed in mass spectrometry of organic compounds.

Nomenclature

Alkyls form homologous series. The simplest series have the general formula C_nH_{2n+1}. Alkyls include *methyl*, CH_3• (named after *methane*), *ethyl* (C_2H_5•), *propyl* (C_3H_7•), *butyl* (C_4H_9•), *pentyl* (C_5H_{11}•), and so on. Alkyl groups that contain one ring have the formula $C_nH2n{-}1$, e.g. cyclopropyl and cyclohexyl. The naming convention is taken from IUPAC nomenclature:

Number of carbon atoms	1	2	3	4	5	6	7	8	9	10	11	12
Prefix	meth-	eth-	prop-	but-	pent-	hex-	hept-	oct-	non-	dec-	undec-	dodec-
Group name	Methyl	Ethyl	Propyl	Butyl	Pentyl	Hexyl	Heptyl	Octyl	Nonyl	Decyl	Undecyl	Dodecyl

These names are used to name branched chained structures, for example 3-methylpentane:

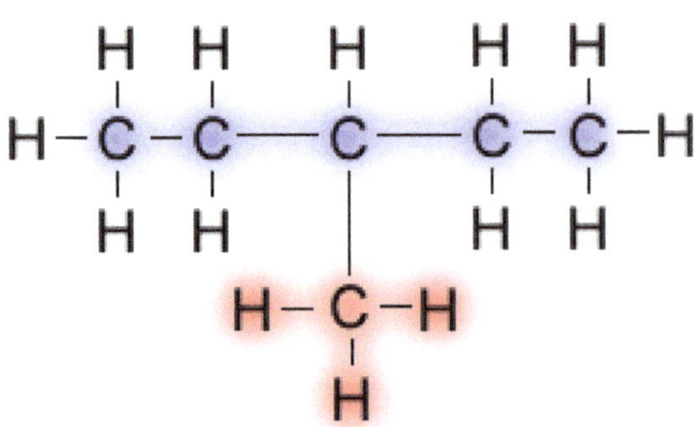

The structure of 3-methylpentane is viewed as consisting of two parts. First, five atoms comprise the longest straight chain of carbon centers. The parent five-carbon compound is named pentane (highlighted blue). The methyl "substituent" or "group" is highlighted red. According to the usual rules of nomenclature, alkyl groups are included in the name of the molecule before the root, as in methylpentane. This name is, however, ambiguous, as the methyl branch could be on various carbon atoms. Thus, the name is 3-methylpentane to avoid ambiguity: The 3- is because the methyl is attached to the third of the five carbon atoms.

If there is more than one of the same alkyl group attached to a chain, then the prefixes are used on the alkyl groups to indicate multiples (i.e., di, tri, tetra, etc.)

In addition, each position on an alkyl chain can be described according to how many other carbon atoms are attached to it. The terms primary, secondary, tertiary, and quaternary refer to a carbon attached to one, two, three, or four other carbons respectively.

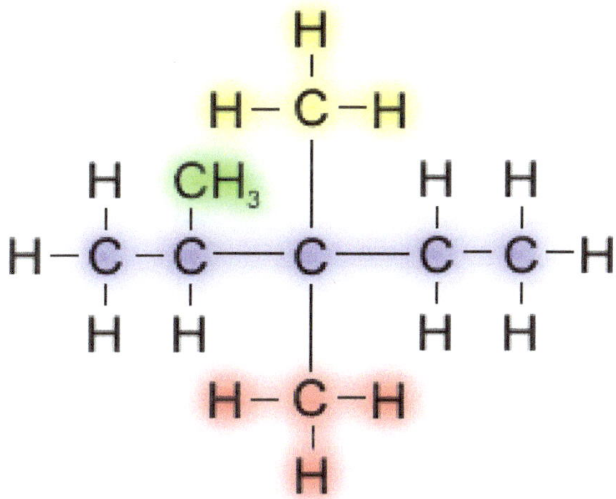

This compound is known as 2,3,3-trimethylpentane. Here three identical alkyl groups attached to carbon atoms 2, 3, and 3. These numbers are included in the name to avoid ambiguity on the position of the groups, and "tri" indicates that there are three identical methyl groups. If one of the methyl groups attached to the third carbon atom were an ethyl group, then the name would become 3-ethyl-2,3 dimethylpentane. Note that when there are different alkyl groups, they are listed in alphabetical order.

Vegetable Oil

Plant oils

Olive oil

A vegetable oil is a triglyceride extracted from a plant. The term "vegetable oil" can be narrowly defined as referring only to plant oils that are liquid at room temperature, or broadly defined without regard to a substance's state of matter at a given temperature. For this reason, vegetable oils that are solid at room temperature are sometimes called vegetable fats. In contrast to these triglycerides, vegetable waxes lack glycerin in their structure. Although many plant parts may yield oil, in commercial practice, oil is extracted primarily from seeds.

On food packaging, the term "vegetable oil" is often used in ingredients lists instead of specifying the exact plant being used, especially when the oil used is less desirable to the consumer or if a mix is used.

Uses of Triglyceride Vegetable Oil

Oils extracted from plants have been used since ancient times and in many cultures. As an example, in a 4,000-year-old kitchen unearthed in Indiana's Charlestown State Park, archaeologist Bob McCullough of Indiana University-Purdue University Fort Wayne found evidence that natives used large slabs of rock to crush hickory nuts, then boiled them in water to extract the oil. Archaeological evidence shows that olives were turned into olive oil by 6000 BC and 4500 BC in present-day Israel.

Culinary Uses

Many vegetable oils are consumed directly, or indirectly as ingredients in food – a role that they share with some animal fats, including butter, ghee, lard, and Schmaltz. The oils serve a number of purposes in this role:

- Shortening – to give pastry a crumbly texture.
- Texture – oils can serve to make other ingredients stick together less.
- Flavor – while less-flavorful oils command premium prices, some oils, such as olive, sesame, or almond oil, may be chosen specifically for the flavor they impart.
- Flavor base – oils can also "carry" flavors of other ingredients, since many flavors are due to chemicals that are soluble in oil.

Secondly, oils can be heated and used to cook other foods. Oils suitable for this objective must have a high flash point. Such oils include the major cooking oils – soybean, rapeseed, canola, sunflower, safflower, peanut, cottonseed, etc. Tropical oils, such as coconut, palm, and rice bran oils, are particularly valued in Asian cultures for high-temperature cooking, because of their unusually high flash points.

Hydrogenated Oils

Unsaturated vegetable oils can be transformed through partial or complete "hydrogenation" into oils of higher melting point. The hydrogenation process involves "sparging" the oil at high temperature and pressure with hydrogen in the presence of a catalyst, typically a powdered nickel compound. As each carbon–carbon double-bond is chemically reduced to a single bond, two hydrogen atoms each form single bonds with the two carbon atoms. The elimination of double bonds by adding hydrogen atoms is called saturation; as the degree of saturation increases, the oil progresses toward being fully hydrogenated. An oil may be hydrogenated to increase resistance to rancidity (oxidation) or to change its physical characteristics. As the degree of saturation increases, the oil's viscosity and melting point increase.

The use of hydrogenated oils in foods has never been completely satisfactory. Because the center arm of the triglyceride is shielded somewhat by the end fatty acids, most of the hydrogenation occurs on the end fatty acids, thus making the resulting fat more brittle. A margarine made from naturally more saturated oils will be more plastic (more "spreadable") than a margarine made from hydrogenated soy oil. While *full hydrogenation* produces largely saturated fatty acids, partial hydrogenation results in the transformation of unsaturated cis fatty acids to unsaturated trans fatty acids in the oil mixture due to the heat used in hydrogenation. Partially hydrogenated oils and their trans fats have been linked to an increased risk of mortality from coronary heart disease, among other increased health risks.

In the US, the Standard of Identity for a product labeled as "vegetable oil margarine" specifies only canola, safflower, sunflower, corn, soybean, or peanut oil may be used. Products not labeled "vegetable oil margarine" do not have that restriction.

Industrial Uses

Vegetable oils are used as an ingredient or component in many manufactured products.

Many vegetable oils are used to make soaps, skin products, candles, perfumes and other personal care and cosmetic products. Some oils are particularly suitable as drying oils, and are used in making paints and other wood treatment products. Dammar oil (a mixture of linseed oil and dammar resin), for example, is used almost exclusively in treating the hulls of wooden boats. Vegetable oils are increasingly being used in the electrical industry as insulators as vegetable oils are not toxic to the environment, biodegradable if spilled and have high flash and fire points. However, vegetable oils are less stable chemically, so they are generally used in systems where they are not exposed to oxygen, and they are more expensive than crude oil distillate. Synthetic tetraesters, which are similar to vegetable oils but with four fatty acid chains compared to the normal three found in a natural ester, are manufactured by Fischer esterification. Tetraesters generally have high stability to oxidation and have found use as engine lubricants. Vegetable oil is being used to produce biodegradable hydraulic fluid and lubricant.

One limiting factor in industrial uses of vegetable oils is that all such oils are susceptible to rancid. Oils that are more stable, such as ben oil or mineral oil, are thus preferred for industrial uses. Castor oil has numerous industrial uses, owing to the presence of hydroxyl group on the fatty acid. Castor oil is a precursor to Nylon 11.

Pet Food Additive

Vegetable oil is used in production of some pet foods. AAFCO defines vegetable oil, in this context, as the product of vegetable origin obtained by extracting the oil from seeds or fruits which are processed for edible purposes. In some poorer-grade pet foods, the oil is listed only as "vegetable oil", without specifying the particular oil.

Fuel

Vegetable oils are also used to make biodiesel, which can be used like conventional diesel. Some vegetable oil blends are used in unmodified vehicles but straight vegetable oil, also known as pure plant oil, needs specially prepared vehicles which have a method of heating the oil to reduce its viscosity. The use of vegetable oils as alternative energy is growing and the availability of biodiesel around the world is increasing.

The NNFCC estimate that the total net greenhouse gas savings when using vegetable oils in place of fossil fuel-based alternatives for fuel production, range from 18 to 100%.

Production

The production process of vegetable oil involves the removal of oil from plant components, typically seeds. This can be done via mechanical extraction using an oil mill or chemical extraction using a solvent. The extracted oil can then be purified and, if required, refined or chemically altered.

Mechanical Extraction

Oils can be removed via mechanical extraction, termed "crushing" or "pressing." This method is typically used to produce the more traditional oils (e.g., olive, coconut etc.), and it is preferred by most "health-food" customers in the United States and in Europe. There are several different types of mechanical extraction. Expeller-pressing extraction is common, though the screw press, ram press, and Ghani (powered mortar and pestle) are also used. Oilseed presses are commonly used in developing countries, among people for whom other extraction methods would be prohibitively expensive; the Ghani is primarily used in India. The amount of oil extracted using these methods varies widely, as shown in the following table for extracting mowrah butter in India:

Method	Percentage extracted
Ghani	20–30%
Expellers	34–37%
Solvent	40–43%

Solvent Extraction

The processing of vegetable oil in commercial applications is commonly done by chemical extraction, using solvent extracts, which produces higher yields and is quicker and less expensive. The most common solvent is petroleum-derived hexane. This technique is used for most of the "newer" industrial oils such as soybean and corn oils.

Supercritical carbon dioxide can be used as a non-toxic alternative to other solvents.

Hydrogenation

Oils may be partially hydrogenated to produce various ingredient oils. Lightly hydrogenated oils have very similar physical characteristics to regular soy oil, but are more resistant to becoming rancid. Margarine oils need to be mostly solid at 32 °C (90 °F) so that the margarine does not melt in warm rooms, yet it needs to be completely liquid at 37 °C (98 °F), so that it doesn't leave a "lardy" taste in the mouth.

Hardening vegetable oil is done by raising a blend of vegetable oil and a catalyst in near-vacuum to very high temperatures, and introducing hydrogen. This causes the carbon atoms of the oil to break double-bonds with other carbons, each carbon forming a new single-bond with a hydrogen atom. Adding these hydrogen atoms to the oil makes it more solid, raises the smoke point, and makes the oil more stable.

Hydrogenated vegetable oils differ in two major ways from other oils which are equally saturated. During hydrogenation, it is easier for hydrogen to come into contact with the fatty acids on the end of the triglyceride, and less easy for them to come into contact with the center fatty acid. This makes the resulting fat more brittle than a tropical oil; soy margarines are less "spreadable". The other difference is that trans fatty acids (often called trans fat) are formed in the hydrogenation reactor, and may amount to as much as 40 percent by weight of a partially hydrogenated oil. Hydrogenated oils, especially partially hydrogenated oils with their higher amounts of trans fatty acids are increasingly thought to be unhealthy.

Deodorization

In the processing of edible oils, the oil is heated under vacuum to near the smoke point, and water is introduced at the bottom of the oil. The water immediately is converted to steam, which bubbles through the oil, carrying with it any chemicals which are water-soluble. The steam sparging removes impurities that can impart unwanted flavors and odors to the oil. Deodorization is key to the manufacture of vegetable oils. Nearly all soybean, corn, and canola oils found on supermarket shelves go through a deodorization stage that removes trace amounts of odors and flavors, and lightens the color of the oil.

Occupational Exposure

People can breathe in vegetable oil mist in the workplace. The U.S. Occupational

Safety and Health Administration (OSHA) has set the legal limit (Permissible exposure limit) for vegetable oil mist exposure in the workplace as 15 mg/m³ total exposure and 5 mg/m³ respiratory exposure over an 8-hour workday. The U.S. National Institute for Occupational Safety and Health (NIOSH) has set a recommended exposure limit (REL) of 10 mg/m³ total exposure and 5 mg/m³ respiratory exposure over an 8-hour workday.

Yield

Some Typical Yields	
Crop	**Yield (MT/ha)**
Palm oil	4.0
Coconut oil	1.4
Canola oil	1.4
Soybean oil	0.6
Sunflower oil	0.6

Particular Oils

The following triglyceride vegetable oils account for almost all worldwide production, by volume. All are used as both cooking oils and as SVO or to make biodiesel. According to the USDA, the total world consumption of major vegetable oils in 2007/08 was:

Oil source	World consumption (million metric tons)	Notes
Palm	41.31	The most widely produced tropical oil, also used to make biofuel
Soybean	41.28	Accounts for about half of worldwide edible oil production
Rapeseed	18.24	One of the most widely used cooking oils, canola is a variety (cultivar) of rapeseed
Sunflower seed	9.91	A common cooking oil, also used to make biodiesel
Peanut	4.82	Mild-flavored cooking oil
Cottonseed	4.99	A major food oil, often used in industrial food processing
Palm kernel	4.85	From the seed of the African palm tree
Coconut	3.48	Used in soaps, cosmetics and cooking

Oil source	World consumption (million metric tons)	Notes
Olive	2.84	Used in cooking, cosmetics, soaps and as a fuel for traditional oil lamps

Note that these figures include industrial and animal feed use. The majority of European rapeseed oil production is used to produce biodiesel, or used directly as fuel in diesel cars which may require modification to heat the oil to reduce its higher viscosity. The suitability of the fuel should come as little surprise, as Rudolf Diesel originally designed his engine to run on peanut oil.

Jambú oil

Other significant triglyceride oils include:

- Corn oil, one of the most common cooking oils. As of 2006 the US produced about 1.09 million metric tons of corn oil, which is used for cooking oil, salad dressing, margarine, mayonnaise, prepared goods like spaghetti sauce and baking mixes, and to fry prepared foods like potato chips and French fries.

Guava oil

- Grape seed oil, used in cooking and cosmetics
- Hazelnut and other nut oils
- Linseed oil, from flax seeds
- Rice bran oil, from rice grains
- Safflower oil, a flavorless and colorless cooking oil
- Sesame oil, used as a cooking oil, and as a massage oil, particularly in India
- Açaí palm oil, used in culinary and cosmetics
- Acmella oleracea - Jambú oil has spilanthol is extracted from the flowers, leaves and stem from jambu
- Annona muricata - Graviola oil is antibacterial, astringent, hypotensive and sedative to name a few
- Astrocaryum aculeatum - Tucumã oil tucuma pulp is used to manufacture soap.
- Brazil nut oil, culinary and cosmetics use
- Carapa oil, pharmaceutical use and anti-mosquito candle
- Mauritia flexuosa - Buriti oil, cosmetics (skin and hair care)
- Passiflora edulis - Passion fruit oil has varied applications in cosmetics manufacturing and for uses as a human or animal food.
- Pentaclethra macroloba - Seje oil, pharmaceutical and cosmetics use

Pracachy or Pentaclethra macroloba

Composition of Fats

Vegetable oils								
Type	Processing Treatment	Saturated fatty acids	Monounsaturated fatty acids	Polyunsaturated fatty acids			Oleic acid (ω-9)	Smoke point
				Total poly	linolenic acid (ω-3)	Linoleic acid (ω-6)		
Avocado		11.560	70.554	13.486	1	12.5		480 °F (249 °C)
Canola (rapeseed)		7.365	63.276	28.142	10	10		400 °F (204 °C)
Coconut		91.000	6.000	3.000		2	6	350 °F (177 °C)
Corn		12.948	27.576	54.677	1	58	28	450 °F (232 °C)
Cottonseed		25.900	17.800	51.900	1	54	19	420 °F (216 °C)
Flaxseed/Linseed (European)		7.500	15.500	79.000	64	15	11	225 °F (107 °C)
Olive		14.000	72.000	14.000	1.5	15		380 °F (193 °C)
Palm		49.300	37.000	9.300		10	40	455 °F (235 °C)
Peanut		16.900	46.200	32.000		32	48	437 °F (225 °C)
Safflower (>70% linoleic)		8.000	15.000	75.000				410 °F (210 °C)
Safflower (high oleic)		7.541	75.221	12.820				410 °F (210 °C)
Soybean		15.650	22.783	57.740	7	50	24	460 °F (238 °C)
Sunflower (<60% linoleic)		10.100	45.400	40.100	0.2	39.8	45.3	440 °F (227 °C)
Sunflower (>70% oleic)		9.859	83.689	3.798				440 °F (227 °C)
Cottonseed (hydrogenated)	Hydrogenated	93.600	1.529	0.587		0.287		
Palm (hydrogenated)	Hydrogenated	47.500	40.600	7.500				
Soybean (hydrogenated)	Hydrogenated	21.100	73.700	0.400	0.096			
Values as percent (%) by weight of total fat.								

History

Such oils have been part of human culture for millennia. Poppy seed,; rapeseed, linseed, almond oil, sesame seed, safflower, and cotton seed were used since at least the bronze age throughout the Middle East and Central Asia. In 1780, Carl Wilhelm Scheele demonstrated that fats were derived from glycerol. Thirty years later Michel Eugène Chevreul deduced that these fats were esters of fatty acids and glycerol.

In modern times, cottonseed oil was marketed by Procter & Gamble as a creamed shortening in 1911. Ginning mills were happy to have someone haul away the cotton seeds. The extracted oil was refined and partially hydrogenated to give a solid at room temperature and thus mimic natural lard, and can it under nitrogen gas. Compared to the rendered lard Procter & Gamble was already selling to consumers, Crisco was cheaper, easier to stir into a recipe, and could be stored at room temperature for two years without turning rancid.

Soybeans were an exciting new crop from China in the 1930s. Soy was protein-rich, and the medium viscosity oil was high in polyunsaturates. Henry Ford established a soybean research laboratory, developed soybean plastics and a soy-based synthetic wool, and built a car "almost entirely" out of soybeans. Roger Drackett had a successful new product with Windex, but he invested heavily in soybean research, seeing it as a smart investment. By the 1950s and 1960s, soybean oil had become the most popular vegetable oil in the US.

In the mid-1970s, Canadian researchers developed a low-erucic-acid rapeseed cultivar. Because the word "rape" was not considered optimal for marketing, they coined the name "canola" (from "Canada Oil low acid"). The U.S. Food and Drug Administration approved use of the canola name in January 1985, and U.S. farmers started planting large areas that spring. Canola oil is lower in saturated fats, and higher in monounsaturates and is a better source of omega-3 fats than other popular oils. Canola is very thin (unlike corn oil) and flavorless (unlike olive oil), so it largely succeeds by displacing soy oil, just as soy oil largely succeeded by displacing cottonseed oil.

Used Oil

A large quantity of used vegetable oil is produced and recycled, mainly from industrial deep fryers in potato processing plants, snack food factories and fast food restaurants.

Recycled oil has numerous uses, including use as a direct fuel, as well as in the production of biodiesel, soap, animal feed, pet food, detergent, and cosmetics. It's traded as the commodity, yellow grease.

Since 2002, an increasing number of European Union countries have prohibited the inclusion of recycled vegetable oil from catering in animal feed. Used cooking oils from food manufacturing, however, as well as fresh or unused cooking oil, continue to be used in animal feed.

Negative Health Effects

Hydrogenated oils have been shown to cause what is commonly termed the "double deadly effect", raising the level of LDLs and decreasing the level of HDLs in the blood, increasing the risk of blood clotting inside blood vessels.

A high consumption of oxidized polyunsaturated fatty acids (PUFAs), which are found in most types of vegetable oil (e.g. soybean oil, corn oil – the most consumed in USA, sunflower oil, etc.) may increase the likelihood that postmenopausal women will develop breast cancer. A similar effect was observed on prostate cancer and skin cancer in mice.

Due to their susceptibility to oxidation from the exposure to oxygen, heat and light, resulting in the formation of oxidation products such as inflammatory peroxides and hydroperoxides, plant oils rich in polyunsatured fatty acids have a limited shelf-life and regular consumption of rancid oils (from prolonged storage or overheating) may have negative health consequences.

Product Labeling

There is increasing concern that the product labeling that includes "vegetable fat" or "vegetable oil" in its list of ingredients masks the identity of the fats or oils present. This has been made more pressing as concerns have been raised over the social and environmental impact of palm oil in particular, especially given the predominance of palm oil.

In Canada, palm oil is one of five vegetable oils, along with palm kernel oil, coconut oil, peanut oil and cocoa butter, which must be specifically named in the list of ingredients for a food product. Also, oils in Canadian food products which have been modified or hydrogenated must contain the word "modified" or "hydrogenated" when listed as an ingredient. A mix of oils other than the aforementioned exceptions may simply be listed as "vegetable oil" in Canada; however, if the food product is a cooking oil, salad oil or table oil, the type of oil must be specified and listing "vegetable oil" as an ingredient is not acceptable.

From 13 December 2014 all food products produced in the European Union will be legally required to indicate the specific vegetable oil used in their manufacture, following the introduction of the Food Information to Consumers Regulation

Diesel Fuel

Diesel fuel in general is any liquid fuel used in diesel engines, whose fuel detonation takes place, without spark, as a result of compression of the inlet air mixture and then injection of fuel. (Glow plugs, grid heaters and heater blocks help achieve high temperatures for combustion during engine startup in cold weather.) Diesel engines have found broad use as a result of higher thermodynamic efficiency and thus fuel efficiency. This is particularly noted where diesel engines are run at part-load; as their air supply is not throttled as in a petrol engine, their efficiency still remains very high.

A tank of diesel fuel on a truck

The most common type of diesel fuel is a specific fractional distillate of petroleum fuel oil, but alternatives that are not derived from petroleum, such as biodiesel, biomass to liquid (BTL) or gas to liquid (GTL) diesel, are increasingly being developed and adopted. To distinguish these types, petroleum-derived diesel is increasingly called petrodiesel. Ultra-low-sulfur diesel (ULSD) is a standard for defining diesel fuel with substantially lowered sulfur contents. As of 2016, almost all of the petroleum-based diesel fuel available in UK, Europe and North America is of a ULSD type. In the UK, diesel fuel for on-road use is commonly abbreviated DERV, standing for *diesel-engined road vehicle*, which carries a tax premium over equivalent fuel for non-road use. In Australia diesel fuel is also known as distillate, and in Indonesia, it is known as Solar, a trademarked name by the local oil company Pertamina.

Origins

Diesel fuel originated from experiments conducted by German scientist and inventor Rudolf Diesel for his compression-ignition engine he invented in 1892. Diesel originally designed his engine to use coal dust as fuel, and experimented with other fuels including vegetable oils such as peanut oil, which was used to power the engines which he exhibited at the 1900 Paris Exposition and the 1911 World's Fair in Paris.

Types

Diesel fuel is produced from various sources, the most common being petroleum. Other sources include biomass, animal fat, biogas, natural gas, and coal liquefaction.

Petroleum Diesel

Petroleum diesel, also called petrodiesel, or fossil diesel is the most common type of diesel fuel. It is produced from the fractional distillation of crude oil between 200 °C (392 °F) and 350 °C (662 °F) at atmospheric pressure, resulting in a mixture of carbon chains that typically contain between 8 and 21 carbon atoms per molecule.

A modern diesel dispenser

Synthetic Diesel

Synthetic diesel can be produced from any carbonaceous material, including biomass, biogas, natural gas, coal and many others. The raw material is gasified into synthesis gas, which after purification is converted by the Fischer–Tropsch process to a synthetic diesel.

The process is typically referred to as biomass-to-liquid (BTL), gas-to-liquid (GTL) or coal-to-liquid (CTL), depending on the raw material used.

Paraffinic synthetic diesel generally has a near-zero content of sulfur and very low aromatics content, reducing unregulated emissions of toxic hydrocarbons, nitrous oxides and particulate matter (PM).

Biodiesel

Fatty-acid methyl ester (FAME), more widely known as biodiesel, is obtained from vegetable oil or animal fats (biolipids) which have been transesterified with methanol. It can be produced from many types of oils, the most common being rapeseed oil (rapeseed methyl ester, RME) in Europe and soybean oil (soy methyl ester, SME) in the US. Methanol can also be replaced with ethanol for the transesterification process, which results in the production of ethyl esters. The transesterification processes use catalysts, such as sodium or potassium hydroxide, to convert vegetable oil and methanol into FAME and the undesirable byproducts glycerine and water, which will need to be removed from the fuel along with methanol traces. FAME can be used pure (B100) in engines where the manufacturer approves such use, but it is more often used as a mix with diesel, BXX where XX is the biodiesel content in percent.

FAME as a fuel is regulated under DIN EN 14214 and ASTM D6751.

FAME has a lower energy content than diesel due to its oxygen content, and as a result, performance and fuel consumption can be affected. It also can have higher levels of NOx emissions, possibly even exceeding the legal limit. FAME also has lower oxidation stability than diesel, and it offers favorable conditions for bacterial growth, so applications which have a low fuel turnover should not use FAME. The loss in power when using pure biodiesel is 5–7%.

Fuel equipment manufacturers (FIE) have raised several concerns regarding FAME fuels: free methanol, dissolved and free water, free glycerin, mono and diglycerides, free fatty acids, total solid impurity levels, alkaline metal compounds in solution and oxidation and thermal stability. They have also identified FAME as being the cause of the following problems: corrosion of fuel injection components, low-pressure fuel system blockage, increased dilution and polymerization of engine sump oil, pump seizures due to high fuel viscosity at low temperature, increased injection pressure, elastomeric seal failures and fuel injector spray blockage.

Unsaturated fatty acids are the source for the lower oxidation stability; they react with oxygen and form peroxides and result in degradation byproducts, which can cause sludge and lacquer in the fuel system.

As FAME contains low levels of sulfur, the emissions of sulfur oxides and sulfates, major components of acid rain, are low. Use of biodiesel also results in reductions of unburned hydrocarbons, carbon monoxide (CO), and particulate matter. CO emissions using biodiesel are substantially reduced, on the order of 50% compared to most petrodiesel fuels. The exhaust emissions of particulate matter from biodiesel have been found to be 30% lower than overall particulate matter emissions from petrodiesel. The exhaust emissions of total hydrocarbons (a contributing factor in the localized formation of smog and ozone) are up to 93% lower for biodiesel than diesel fuel.

Biodiesel also may reduce health risks associated with petroleum diesel. Biodiesel emissions showed decreased levels of polycyclic aromatic hydrocarbon (PAH) and nitrited PAH compounds, which have been identified as potential cancer-causing compounds. In recent testing, PAH compounds were reduced by 75–85%, except for benz(a)anthracene, which was reduced by roughly 50%. Targeted nPAH compounds were also reduced dramatically with biodiesel fuel, with 2-nitrofluorene and 1-nitropyrene reduced by 90%, and the rest of the nPAH compounds reduced to only trace levels.

Hydrogenated Oils and Fats

This category of diesel fuels involves converting the triglycerides in vegetable oil and animal fats into alkanes by refining and hydrogenation, such as H-Bio. The produced fuel has many properties that are similar to synthetic diesel, and are free from the many disadvantages of FAME.

DME

Dimethyl ether, DME, is a synthetic, gaseous diesel fuel that results in clean combustion with very little soot and reduced NOx emissions.

Storage

In the US, diesel is recommended to be stored in a yellow container to differentiate it from kerosene and gasoline, which are typically kept in blue and red containers, respectively. In the UK, diesel is normally stored in a black container, to differentiate it from unleaded petrol (which is commonly stored in a green container) and leaded petrol (which is stored in a red container).

Measurements and Pricing

Cetane Number

The principal measure of diesel fuel quality is its cetane number. A cetane number is a measure of the delay of ignition of a diesel fuel. A higher cetane number indicates that the fuel ignites more readily when sprayed into hot compressed air. European (EN 590 standard) road diesel has a minimum cetane number of 51. Fuels with higher cetane numbers, normally "premium" diesel fuels with additional cleaning agents and some synthetic content, are available in some markets.

Fuel Value and Price

As of 2010, the density of petroleum diesel is about 0.832 kg/L (6.943 lb/US gal), about 11.6% more than ethanol-free petrol (gasoline), which has a density of about 0.745 kg/L (6.217 lb/US gal). About 86.1% of the fuel mass is carbon, and when burned, it offers a net heating value of 43.1 MJ/kg as opposed to 43.2 MJ/kg for gasoline. However, due to the higher density, diesel offers a higher volumetric energy density at 35.86 MJ/L (128,700 BTU/US gal) vs. 32.18 MJ/L (115,500 BTU/US gal) for gasoline, some 11% higher, which should be considered when comparing the fuel efficiency by volume. The CO_2 emissions from diesel are 73.25 g/MJ, just slightly lower than for gasoline at 73.38 g/MJ. Diesel is generally simpler to refine from petroleum than gasoline, and contains hydrocarbons having a boiling point in the range of 180–360 °C (360–680 °F). The price of diesel traditionally rises during colder months as demand for heating oil rises, which is refined in much the same way. Because of recent changes in fuel quality regulations, additional refining is required to remove sulfur, which contributes to a sometimes higher cost. In many parts of the United States and throughout the United Kingdom and Australia, diesel may be priced higher than petrol. Reasons for higher-priced diesel include the shutdown of some refineries in the Gulf of Mexico, diversion of mass refining capacity to gasoline production, and a recent transfer to ultra-low-sulfur diesel (ULSD), which causes infrastructural complications. In Sweden, a diesel fuel designated as MK-1 (class 1 environmental diesel) is also being sold; this is a ULSD that also has a lower aromatics content, with a limit of 5%. This fuel is slightly more expensive to produce than regular ULSD.

Taxation

Diesel fuel is very similar to heating oil, which is used in central heating. In Europe, the United States, and Canada, taxes on diesel fuel are higher than on heating oil due to the fuel tax, and in those areas, heating oil is marked with fuel dyes and trace chemicals to prevent and detect tax fraud. "Untaxed" diesel (sometimes called "off-road diesel") is available in some countries for use primarily in agricultural applications, such as fuel for tractors, recreational and utility vehicles or other noncommercial vehicles that do not use public roads. This fuel may have sulfur levels that exceed the limits for road use in some countries (e.g. US).

This untaxed diesel is dyed red for identification, and using this untaxed diesel fuel for a typically taxed purpose (such as driving use), the user can be fined (e.g. US$10,000 in the US). In the United Kingdom, Belgium and the Netherlands, it is known as red diesel (or gas oil), and is also used in agricultural vehicles, home heating tanks, refrigeration units on vans/trucks which contain perishable items such as food and medicine and for marine craft. Diesel fuel, or marked gas oil is dyed green in the Republic of Ireland and Norway. The term "diesel-engined road vehicle" (DERV) is used in the UK as a synonym for unmarked road diesel fuel. In India, taxes on diesel fuel are lower than on petrol, as the majority of the transportation for grain and other essential commodities across the country runs on diesel.

Taxes on biodiesel in the US vary between states; some states (Texas, for example) have no tax on biodiesel and a reduced tax on biodiesel blends equivalent to the amount of biodiesel in the blend, so that B20 fuel is taxed 20% less than pure petrodiesel. Other states, such as North Carolina, tax biodiesel (in any blended configuration) the same as petrodiesel, although they have introduced new incentives to producers and users of all biofuels.

Uses

Unlike gasoline and liquefied petroleum gas engines, diesel engines do not use high-voltage spark ignition (spark plugs). An engine running on diesel compresses the air inside the cylinder to high pressures and temperatures (compression ratios from 14:1 to 18:1 are common in current diesel engines); the engine generally injects the diesel fuel directly into the cylinder, starting a few degrees before top dead center (TDC) and continuing during the combustion event. The high temperatures inside the cylinder cause the diesel fuel to react with the oxygen in the mix (burn or oxidize), heating and expanding the burning mixture to convert the thermal/pressure difference into mechanical work, i.e., to move the piston.

Engines have glow plugs and grid heaters to help start the engine by preheating the cylinders to a minimum operating temperature. Diesel engines are lean burn engines, burning the fuel in more air than is needed for the chemical reaction. They thus use less fuel than rich burn spark ignition engines which use a stoichiometric air-fuel ratio (just enough air to react with the fuel). As Professor Harvey of the University of Toronto notes, "due to the absence of throttling [constant amount of air admitted, per unit fuel, with no user-determined variation], and the high compression ratio and lean fuel mixture, diesel engines are substantially more efficient than spark-ignited engines", generally; Harvey cites the side-by-side comparisons of Schipper et al. and the estimates of >20% lower fuel use and (given difference in differences in energy content between fuel types) >15% lower energy use. Gas turbine and some other types of internal combustion engines, and external combustion engine, both can also be designed to take diesel fuel.

The viscosity requirement of diesel fuel is usually specified at 40 °C. A disadvantage of diesel as a vehicle fuel in cold climates, is that its viscosity increases as the temperature decreases, changing it into a gel that cannot flow in fuel systems. Special low-temperature diesel contains additives to keep it liquid at lower temperatures, but starting a diesel engine in very cold weather may still pose considerable difficulties. Another disadvantage of diesel engines compared to petrol/gasoline engines is the possibility of diesel engine runaway failure. Since diesel engines do not need spark ignition, they can run as long as diesel fuel is supplied. Fuel is typically supplied via a fuel pump. If the pump breaks down in an "open" position, the supply of fuel will be unrestricted, and the engine will run away and risk terminal failure.

With turbocharged engines, the oil seals on the turbocharger may fail, allowing lubricating oil into the combustion chamber, where it is burned like regular diesel fuel. In vehicles or installations that use diesel engines and also bottled gas, a gas leak into the engine room could also provide fuel for a runaway, via the engine air intake.

The crank case ventilation of modern road-use diesel engines is diverted into the intake manifold, because ventilating the crank case into outside air is inadvisable due to lubricant mist it contains. If the engine's piston rings malfunction, this will cause excessive pressure in the crank case forcing mist of engine lubricant into the intake manifold. Since most engines use oil which can be burnt in the same fashion as diesel, this will result in diesel engine runaway. To prevent that, more premium crank case ventilation solutions are fitted with a filter to catch out lubricant mist.

Most modern road use diesel engines are provided with an FRP valve in the intake manifold (usually mistaken by some as a petrol engine throttle body). In most basic applications this valve will close a flow of air mixture to the engine when the vehicle is switched off, preventing diesel engine runaway by starving the engine of oxygen; this will also make standard shutdown much smoother by eliminating compression and decompression rattle by making the pistons effectively work in vacuum. In more advanced control systems this FRP valve can be shut by an electronic control unit when it senses runaway scenario.

Trucks

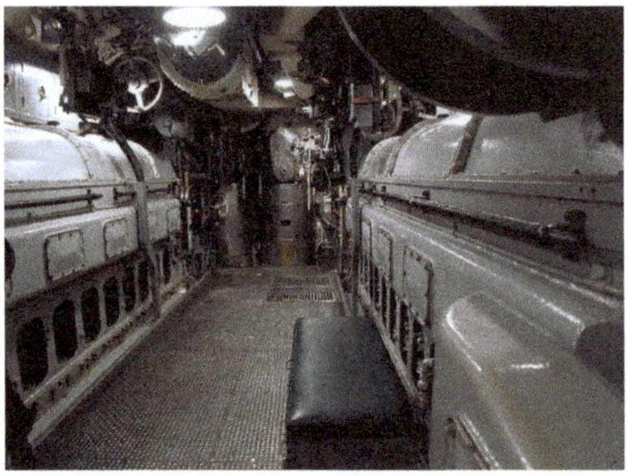

Fairbanks-Morse opposed piston diesel engines on a submarine

Diesel fuel is widely used in most types of transportation. Trucks and buses, which were often gasoline-powered in the 1920s through 1950s, are now almost exclusively diesel-powered. The gasoline-powered passenger automobile is the major exception; diesel cars are less numerous worldwide.

Railroad

Diesel displaced coal and fuel oil for steam-powered vehicles in the latter half of the 20th century, and is now used almost exclusively for the combustion engines of self-powered rail vehicles (locomotives and railcars).

Aircraft

Packard DR-980 9-cylinder diesel aircraft engine

The first diesel-powered flight of a fixed-wing aircraft took place on the evening of 18 September 1928, at the Packard Proving Grounds near Utica, Michigan. With Captain Lionel M. Woolson and Walter Lees at the controls the first "official" test flight was taken the next morning, flying a Stinson SM1B (X7654), powered by a Packard DR-980 9-cylinder diesel radial engine, designed by Woolson. Charles Lindbergh flew the same aircraft and in 1929, it was flown 621 miles (999 km) nonstop from Detroit to Langley Field, near Norfolk, Virginia. In 1931, Walter Lees and Fredrick Brossy set the nonstop flight record flying a Bellanca powered by a Packard diesel for 84 hours and 32 minutes. *X7654* is now owned by Greg Herrick and is at the Golden Wings Flying Museum near Minneapolis, Minnesota.

Diesel engines for airships were developed in both Germany and the United Kingdom by Daimler-Benz and Beardmore produced the Daimler-Benz DB 602 and Beardmore Typhoon respectively. The LZ 129 Hindenburg rigid airship was powered by four Daimler-Benz DB 602 16-cylinder diesel engines, each with 1,200 hp (890 kW) available in bursts and 850 horsepower (630 kW) available for cruising. The Beardmore Typhoon powered the ill-fated R101 airship, built for the Empire airship programme in 1931.

With a production run of at least 900 engines, the most-produced aviation diesel engine in history was probably the Junkers Jumo 205. Similar developments from the Junkers Motorenwerke and licence-built versions of the Jumo 204 and Jumo 205, boosted German diesel aero-engine production to at least 1000 examples, the vast majority of which were liquid-cooled, opposed-piston, two-stroke engines.

In the Soviet Union significant progress towards practical diesel aero-engines was made by the TsIAM (*Tsentral'nyy Institut Aviatsionnovo Motorostroyeniya* – central institute of aviation motors) and particularly by A.D. Charomskiy, who nursed the Charomskiy ACh-30 into production and limited operational use.

Military Vehicles

Armored fighting vehicles use diesel because of its lower flammability risks and the engines' higher provision of torque and lower likelihood of stalling.

Cars

Diesel-powered cars generally have a better fuel economy than equivalent gasoline engines and produce less greenhouse gas emission. Their greater economy is due to the higher energy per-litre content of diesel fuel and the intrinsic efficiency of the diesel engine. While petrodiesel's higher density results in higher greenhouse gas emissions per litre compared to gasoline, the 20–40% better fuel economy achieved by modern diesel-engined automobiles offsets the higher per-litre emissions of greenhouse gases, and a diesel-powered vehicle emits 10–20 percent less greenhouse gas than comparable gasoline vehicles. Biodiesel-powered diesel engines offer substantially improved emission reductions compared to petrodiesel or gasoline-powered engines, while retaining most of the fuel economy advantages over conventional gasoline-powered automobiles. However, the increased compression ratios mean there are increased emissions of oxides of nitrogen (NO_x) from diesel engines. This is compounded by biological nitrogen in biodiesel to make NO_x emissions the main drawback of diesel versus gasoline engines.

Tractors and Heavy Equipment

Today's tractors and heavy equipment are mostly diesel-powered. Among tractors, only the smaller classes may also offer gasoline engines. The dieselization of tractors and heavy equipment began in Germany before World War II but was unusual in the United States until after that war. During the 1950s and 1960s, it progressed in the latter country as well.

Tractors and heavy equipment were often multifuel in the 1920s through 1940s, running spark-ignition and low-compression engines. Thus many farm tractors of the era could burn gasoline, alcohol, kerosene, and any light grade of fuel oil such as diesel fuel, heating oil, or tractor vaporising oil, according to whichever was most affordable in any region at any given time. On U.S. farms during this era, the name "distillate" often referred to any of the aforementioned light fuel oils. The engines did not start as well on distillate, so typically a small auxiliary gasoline tank was used for cold starting, and the fuel valves were adjusted several minutes later, after warm-up, to switch to distillate. Engine accessories such as vaporizers and radiator shrouds were also used, both with the aim of capturing heat, because when such an engine was run on distillate, it ran better when both it and the air it inhaled were warmer rather than at ambient temperature. Dieselization with dedicated diesel engines (high-compression with mechanical fuel injection and compression ignition) replaced such systems and made more efficient use of the diesel fuel being burned.

Other Uses

Poor quality diesel fuel has been used as an extraction agent for liquid–liquid extraction of palladium from nitric acid mixtures. Such use has been proposed as a means of separating the fission product palladium from PUREX raffinate which comes from used nuclear fuel. In this system of solvent extraction, the hydrocarbons of the diesel act as the diluent while the dialkyl sulfides act as the extractant. This extraction operates by a solvation mechanism. So far, neither a pilot plant nor full scale plant has been constructed to recover palladium, rhodium or ruthenium from nuclear wastes created by the use of nuclear fuel.

Diesel fuel is also often used as the main ingredient in oil-base mud drilling fluid. The advantage of using diesel is its low cost and that it delivers excellent results when drilling a wide variety of difficult strata including shale, salt and gypsum formations. Diesel-oil mud is typically mixed with up to 40% brine water. Due to health, safety and environmental concerns, Diesel-oil mud is often replaced with vegetable, mineral, or synthetic food-grade oil-base drilling fluids, although diesel-oil mud is still in widespread use in certain regions.

During development of rocket engines in Germany during World War II J-2 Diesel fuel was used as the fuel component in several engines including the BMW 109-718. J-2 diesel fuel was also used as a fuel for gas turbine engines.

Chemical Analysis

Chemical Composition

Diesel does not mix with water.

Petroleum-derived diesel is composed of about 75% saturated hydrocarbons (primarily paraffins including *n*, *iso*, and cycloparaffins), and 25% aromatic hydrocarbons (including naphthalenes and alkylbenzenes). The average chemical formula for common diesel fuel is $C_{12}H_{23}$, ranging approximately from $C_{10}H_{20}$ to $C_{15}H_{28}$.

Chemical Properties

Most diesel fuels freeze at common winter temperatures, while the temperatures greatly vary. Petrodiesel typically freezes around temperatures of −8.1 °C (17.5 °F), whereas biodiesel freezes between temperatures of 2° to 15 °C (35° to 60 °F). The viscosity of diesel noticeably increases as the temperature decreases, changing it into a gel at temperatures of −19 °C (−2.2 °F) to −15 °C (5 °F), that cannot flow in fuel systems. Conventional diesel fuels vaporise at temperatures between 149 °C and 371 °C.

Conventional diesel flash points vary between 52 and 96 °C, which makes it safer than petrol and unsuitable for spark-ignition engines. Unlike petrol, the flash point of a diesel fuel has no relation to its performance in an engine nor to its auto ignition qualities.

Hazards

Reduction of Sulfur Emissions

In the past, diesel fuel contained higher quantities of sulfur. European emission standards and preferential taxation have forced oil refineries to dramatically reduce the level of sulfur in diesel fuels. In the European Union, the sulfur content has dramatically reduced during the last 20 years. Automotive diesel fuel is covered in the European Union by standard EN 590. In the 1990s specifications allowed a content of 2000 ppm max of sulphur, reduced to a limit of 350 ppm by the beginning of the 21st century with the introduction of Euro 3 specifications. The limit was lowered with the introduction of Euro 4 by 2006 to 50 ppm (ULSD, Ultra Low Sulfur Diesel). The standard currently in force in European Europe for Diesel Fuel is the Euro 5, with a maximum content of 10 ppm.

Emission standard	At latest	Sulfur content	Cetane number
Euro 1	1 January 1993	max. 2000 ppm	min. 49
Euro 2	1 January 1996	max. 500 ppm	min. 49
Euro 3	1 January 2001	max. 350 ppm	min. 51
Euro 4	1. January 2006	max. 50 ppm	min. 51
Euro 5	1 January 2009	max. 10 ppm	min. 51

In the United States, more stringent emission standards have been adopted with the transition to ULSD starting in 2006, and becoming mandatory on June 1, 2010. U.S. diesel fuel typically also has a lower cetane number (a measure of ignition quality) than European diesel, resulting in worse cold weather performance and some increase in emissions.

Environment Hazards of Sulfur

High levels of sulfur in diesel are harmful for the environment because they prevent the use of catalytic diesel particulate filters to control diesel particulate emissions, as well as more advanced technologies, such as nitrogen oxide (NO_x) adsorbers (still under development), to reduce emissions. Moreover, sulfur in the fuel is oxidized during combustion, producing sulfur dioxide and sulfur trioxide, that in presence of water rapidly convert to sulfuric acid, one of the chemical processes that results in acid rain. However, the process for lowering sulfur also reduces the lubricity of the fuel, meaning that additives must be put into the fuel to help lubricate engines. Biodiesel and biodiesel/petrodiesel blends, with their higher lubricity levels, are increasingly being utilized as an alternative. The U.S. annual consumption of diesel fuel in 2006 was about 190 billion litres (42 billion imperial gallons or 50 billion US gallons).

Algae, Microbes, and Water Contamination

There has been much discussion and misunderstanding of algae in diesel fuel. Algae need light to live and grow. As there is no sunlight in a closed fuel tank, no algae can survive, but some microbes can survive and feed on the diesel fuel.

These microbes form a colony that lives at the interface of fuel and water. They grow quite fast in warmer temperatures. They can even grow in cold weather when fuel tank heaters are installed. Parts of the colony can break off and clog the fuel lines and fuel filters.

Water in fuel can damage a fuel injection pump; some diesel fuel filters also trap water. Water contamination in diesel fuel can lead to freezing while in the fuel tank. The freezing water that saturates the fuel will sometimes clog the fuel injector pump. Once the water inside the fuel tank has started to freeze, gelling is more likely to occur. When the fuel is gelled it is not effective until the temperature is raised and the fuel returns to a liquid state.

Road hazard

Petrodiesel spilled on a roadway poses a hazard to vehicles, due to its high evaporation temperature. After the light fractions have evaporated, a greasy slick is left on the road which can destabilize moving vehicles, causing them to skid. Diesel spills severely reduce tire grip and traction, and have been implicated in many accidents. The loss of traction is similar to that encountered on black ice. Diesel slicks are especially dangerous for two-wheeled vehicles such as motorcycles.

References

- Tillotson, Geoffrey (1981). "Engines for Main Battle Tanks". In Col. John Weeks. Jane's 1981–82 Military Annual. Jane's. p. 59,63. ISBN 0-7106-0137-9.
- Duffy, Michael C. (1 January 2003). Electric Railways 1880–1990. London: Institution of Engineering and Technology. ISBN 0-85296-805-1. Retrieved 9 October 2014.
- Wellington, B.F.; Asmus, Alan F. (1995). Diesel Engines and Fuel Systems. Longman Australia. ISBN 0-582-90987-2.
- Solomon, Brian; Yough, Patrick (15 July 2009). Coal Trains: The History of Railroading and Coal in the United States (Google eBook). MBI Publishing Company. ISBN 0-7603-3359-9. Retrieved 9 October 2014.
- Kotelnikov, Vladimir (2005). Russian Piston Aero Engines. Marlborough: The Crowood Press Ltd. ISBN 978-1-86126-702-3.
- Ayhan Demirbas (2008). Biodiesel: A Realistic Fuel Alternative for Diesel Engines. Berlin: Springer. p. 74. ISBN 1-84628-994-7.
- Chris Collins (2007), "Implementing Phytoremediation of Petroleum Hydrocarbons, Methods in Biotechnology 23:99–108. Humana Press. ISBN 1-58829-541-9.
- Hiereth, Hermann; Drexl, Klaus; Prenninger, Peter (4 November 2007). Charging the Internal Combustion Engine. Springer Science & Business Media. p. 233. ISBN 978-0-471-55878-1. Retrieved 28 July 2014.
- Date, Anil W. (7 March 2011). Analytic Combustion: With Thermodynamics, Chemical Kinetics and Mass Transfer (Google eBook). Cambridge University Press. ISBN 1-107-00286-9. Retrieved 9 October 2014.
- "Basic Labelling Requirements", Guide to Food Labelling and Advertising, Canadian Food Inspection Agency, retrieved 2015-04-08

2

Biofuel: An Integrated Study

Fuel can be produced by geological processes or by biological processes. Biofuel is produced by biological processes and is derived directly from plants or indirectly from agriculture or domestic/industrial waste. The major biofuels discussed in the following text are cellulosic ethanol, vegetable oil fuel, ethanol fuel, babassu oil, methanol fuel and biogas. This section will provide an integrated understanding of biofuel.

Biofuel

A biofuel is a fuel that is produced through contemporary biological processes, such as agriculture and anaerobic digestion, rather than a fuel produced by geological processes such as those involved in the formation of fossil fuels, such as coal and petroleum, from prehistoric biological matter. Biofuels can be derived directly from plants, or indirectly from agricultural, commercial, domestic, and/or industrial wastes. Renewable biofuels generally involve contemporary carbon fixation, such as those that occur in plants or microalgae through the process of photosynthesis. Other renewable biofuels are made through the use or conversion of biomass (referring to recently living organisms, most often referring to plants or plant-derived materials). This biomass can be converted to convenient energy-containing substances in three different ways: thermal conversion, chemical conversion, and biochemical conversion. This biomass conversion can result in fuel in solid, liquid, or gas form. This new biomass can also be used directly for biofuels.

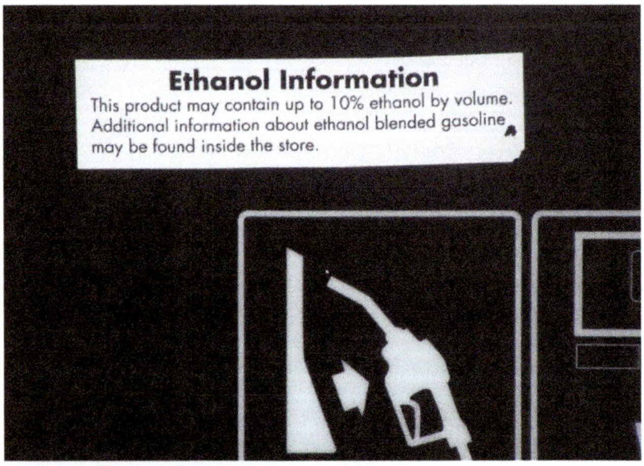

Information on pump regarding ethanol fuel blend up to 10%, California

Bioethanol is an alcohol made by fermentation, mostly from carbohydrates produced in sugar or starch crops such as corn, sugarcane, or sweet sorghum. Cellulosic biomass, derived from non-food sources, such as trees and grasses, is also being developed as a feedstock for ethanol production. Ethanol can be used as a fuel for vehicles in its pure form, but it is usually used as a gasoline additive to increase octane and improve vehicle emissions. Bioethanol is widely used in the USA and in Brazil. Current plant design does not provide for converting the lignin portion of plant raw materials to fuel components by fermentation.

Biodiesel can be used as a fuel for vehicles in its pure form, but it is usually used as a diesel additive to reduce levels of particulates, carbon monoxide, and hydrocarbons from diesel-powered vehicles. Biodiesel is produced from oils or fats using transesterification and is the most common biofuel in Europe.

In 2010, worldwide biofuel production reached 105 billion liters (28 billion gallons US), up 17% from 2009, and biofuels provided 2.7% of the world's fuels for road transport. Global ethanol fuel production reached 86 billion liters (23 billion gallons US) in 2010, with the United States and Brazil as the world's top producers, accounting together for 90% of global production. The world's largest biodiesel producer is the European Union, accounting for 53% of all biodiesel production in 2010. As of 2011, mandates for blending biofuels exist in 31 countries at the national level and in 29 states or provinces. The International Energy Agency has a goal for biofuels to meet more than a quarter of world demand for transportation fuels by 2050 to reduce dependence on petroleum and coal. The production of biofuels also led into a flourishing automotive industry, where by 2010, 79% of all cars produced in Brazil were made with a hybrid fuel system of bioethanol and gasoline.

There are various social, economic, environmental and technical issues relating to biofuels production and use, which have been debated in the popular media and scientific journals. These include: the effect of moderating oil prices, the "food vs fuel" debate, poverty reduction potential, carbon emissions levels, sustainable biofuel production, deforestation and soil erosion, loss of biodiversity, impact on water resources, rural social exclusion and injustice, shantytown migration, rural unskilled unemployment, and nitrous oxide (NO_2) emissions.

Liquid Fuels for Transportation

Most transportation fuels are liquids, because vehicles usually require high energy density. This occurs naturally in liquids and solids. High energy density can also be provided by an internal combustion engine. These engines require clean-burning fuels. The fuels that are easiest to burn cleanly are typically liquids and gases. Thus, liquids meet the requirements of being both energy-dense and clean-burning. In addition, liquids (and gases) can be pumped, which means handling is easily mechanized, and thus less laborious.

First-generation Biofuels

"First-generation" or conventional biofuels are made from sugar, starch, or vegetable oil.

Ethanol

Neat ethanol on the left (A), gasoline on the right (G) at a filling station in Brazil

Biologically produced alcohols, most commonly ethanol, and less commonly propanol and butanol, are produced by the action of microorganisms and enzymes through the fermentation of sugars or starches (easiest), or cellulose (which is more difficult). Biobutanol (also called biogasoline) is often claimed to provide a direct replacement for gasoline, because it can be used directly in a gasoline engine.

Ethanol fuel is the most common biofuel worldwide, particularly in Brazil. Alcohol fuels are produced by fermentation of sugars derived from wheat, corn, sugar beets, sugar cane, molasses and any sugar or starch from which alcoholic beverages such as whiskey, can be made (such as potato and fruit waste, etc.). The ethanol production methods used are enzyme digestion (to release sugars from stored starches), fermentation of the sugars, distillation and drying. The distillation process requires significant energy input for heat (sometimes unsustainable natural gas fossil fuel, but cellulosic biomass such as bagasse, the waste left after sugar cane is pressed to extract its juice, is the most common fuel in Brazil, while pellets, wood chips and also waste heat are more common in Europe) Waste steam fuels ethanol factory- where waste heat from the factories also is used in the district heating grid.

Ethanol can be used in petrol engines as a replacement for gasoline; it can be mixed with gasoline to any percentage. Most existing car petrol engines can run on blends of up to 15% bioethanol with petroleum/gasoline. Ethanol has a smaller energy density than that of gasoline; this means it takes more fuel (volume and mass) to produce the same amount of work. An advantage of ethanol (CH_3CH_2OH) is that it has a higher octane rating than ethanol-free gasoline available at roadside gas stations, which allows an increase of an engine's compression ratio for increased thermal efficiency. In high-altitude (thin air) locations, some states mandate a mix of gasoline and ethanol as a winter oxidizer to reduce atmospheric pollution emissions.

Ethanol is also used to fuel bioethanol fireplaces. As they do not require a chimney and are "flueless", bioethanol fires are extremely useful for newly built homes and apartments without a flue. The downsides to these fireplaces is that their heat output is slightly less than electric heat or gas fires, and precautions must be taken to avoid carbon monoxide poisoning.

U.S. President George W. Bush looks at sugar cane, a source of biofuel, with Brazilian President Luiz Inácio Lula da Silva during a tour on biofuel technology at Petrobras in São Paulo, Brazil, 9 March 2007.

Corn-to-ethanol and other food stocks has led to the development of cellulosic ethanol. According to a joint research agenda conducted through the US Department of Energy, the fossil energy ratios (FER) for cellulosic ethanol, corn ethanol, and gasoline are 10.3, 1.36, and 0.81, respectively.

Ethanol has roughly one-third lower energy content per unit of volume compared to gasoline. This is partly counteracted by the better efficiency when using ethanol (in a long-term test of more than 2.1 million km, the BEST project found FFV vehicles to be 1-26 % more energy efficient than petrol cars The BEST project), but the volumetric consumption increases by approximately 30%, so more fuel stops are required.

With current subsidies, ethanol fuel is slightly cheaper per distance traveled in the United States.

Biodiesel

Biodiesel is the most common biofuel in Europe. It is produced from oils or fats using transesterification and is a liquid similar in composition to fossil/mineral diesel. Chemically, it consists mostly of fatty acid methyl (or ethyl) esters (FAMEs). Feedstocks for biodiesel include animal fats, vegetable oils, soy, rapeseed, jatropha, mahua, mustard, flax, sunflower, palm oil, hemp, field pennycress, *Pongamia pinnata* and algae. Pure biodiesel (B100) currently reduces emissions with up to 60% compared to diesel Second generation B100.

Biodiesel can be used in any diesel engine when mixed with mineral diesel. In some countries, manufacturers cover their diesel engines under warranty for B100 use, although Volkswagen of Germany, for example, asks drivers to check by telephone with the VW environmental services department before switching to B100. B100 may become more viscous at lower temperatures, depending on the feedstock used. In most cases, biodiesel is compatible with diesel engines from 1994 onwards, which use 'Viton' (by DuPont) synthetic rubber in their mechanical fuel injection systems. Note however, that no vehicles are certified for using neat biodiesel before 2014, as there was no emission control protocol available for biodiesel before this date.

Electronically controlled 'common rail' and 'unit injector' type systems from the late 1990s onwards may only use biodiesel blended with conventional diesel fuel. These engines have finely metered and atomized multiple-stage injection systems that are very sensitive to the viscosity of the fuel. Many current-generation diesel engines are made so that they can run on B100 without altering the engine itself, although this depends on the fuel rail design. Since biodiesel is an effective solvent and cleans residues deposited by mineral diesel, engine filters may need to be replaced more often, as the biofuel dissolves old deposits in the fuel tank and pipes. It also effectively cleans the engine combustion chamber of carbon deposits, helping to maintain efficiency. In many European countries, a 5% biodiesel blend is widely used and is available at thousands of gas stations. Biodiesel is also an oxygenated fuel, meaning it contains a reduced amount of carbon and higher hydrogen and oxygen content than fossil diesel. This improves the combustion of biodiesel and reduces the particulate emissions from unburnt carbon. However, using neat biodiesel may increase NOx-emissions Nylund.N-O & Koponen.K. 2013. Fuel and Technology Alternatives for Buses. Overall Energy Efficiency and Emission Performance. IEA Bioenergy Task 46. Possibly the new emission standards Euro VI/EPA 10 will lead to reduced NOx-levels also when using B100.

Biodiesel is also safe to handle and transport because it is non-toxic and biodegradable, and has a high flash point of about 300 °F (148 °C) compared to petroleum diesel fuel, which has a flash point of 125 °F (52 °C).

In the USA, more than 80% of commercial trucks and city buses run on diesel. The emerging US biodiesel market is estimated to have grown 200% from 2004 to 2005. "By the end of 2006 biodiesel production was estimated to increase fourfold [from 2004] to more than" 1 billion US gallons (3,800,000 m^3).

In France, biodiesel is incorporated at a rate of 8% in the fuel used by all French diesel vehicles. Avril Group produces under the brand Diester, a fifth of 11 million tons of biodiesel consumed annually by the European Union. It is the leading European producer of biodiesel.

Other Bioalcohols

Methanol is currently produced from natural gas, a non-renewable fossil fuel. In the future it is hoped to be produced from biomass as biomethanol. This is technically feasible, but the production is currently being postponed for concerns of Jacob S. Gibbs and Brinsley Coleberd that the economic viability is still pending. The methanol economy is an alternative to the hydrogen economy, compared to today's hydrogen production from natural gas.

Butanol (C_4H_9OH) is formed by ABE fermentation (acetone, butanol, ethanol) and experimental modifications of the process show potentially high net energy gains with butanol as the only liquid product. Butanol will produce more energy and allegedly can be burned "straight" in existing gasoline engines (without modification to the engine or car), and is less corrosive and less water-soluble than ethanol, and could be distributed via existing infrastructures. DuPont and BP are working together to help develop butanol. *E. coli* strains have also been successfully engineered to produce butanol by modifying their amino acid metabolism.

Green Diesel

Green diesel is produced through hydrocracking biological oil feedstocks, such as vegetable oils and animal fats. Hydrocracking is a refinery method that uses elevated temperatures and pressure in the presence of a catalyst to break down larger molecules, such as those found in vegetable oils, into shorter hydrocarbon chains used in diesel engines. It may also be called renewable diesel, hydrotreated vegetable oil or hydrogen-derived renewable diesel. Green diesel has the same chemical properties as petroleum-based diesel. It does not require new engines, pipelines or infrastructure to distribute and use, but has not been produced at a cost that is competitive with petroleum. Gasoline versions are also being developed. Green diesel is being developed in Louisiana and Singapore by ConocoPhillips, Neste Oil, Valero, Dynamic Fuels, and Honeywell UOP as well as Preem in Gothenburg, Sweden, creating what is known as Evolution Diesel.

Biofuel Gasoline

In 2013 UK researchers developed a genetically modified strain of Escherichia coli (E.Coli), which could transform glucose into biofuel gasoline that does not need to be blended. Later in 2013 UCLA researchers engineered a new metabolic pathway to bypass glycolysis and increase the rate of conversion of sugars into biofuel, while KAIST researchers developed a strain capable of producing short-chain alkanes, free fatty acids, fatty esters and fatty alcohols through the fatty acyl (acyl carrier protein (ACP)) to fatty acid to fatty acyl-CoA pathway *in vivo*. It is believed that in the future it will be possible to "tweak" the genes to make gasoline from straw or animal manure.

Vegetable Oil

Filtered waste vegetable oil

Straight unmodified edible vegetable oil is generally not used as fuel, but lower-quality oil can and has been used for this purpose. Used vegetable oil is increasingly being processed into biodiesel, or (more rarely) cleaned of water and particulates and used as a fuel.

As with 100% biodiesel (B100), to ensure the fuel injectors atomize the vegetable oil in the correct pattern for efficient combustion, vegetable oil fuel must be heated to reduce its viscosity to that of diesel, either by electric coils or heat exchangers. This is easier in warm or temperate climates. MAN B&W Diesel, Wärtsilä, and Deutz AG, as well as a number of smaller companies, such as Elsbett, offer engines that are compatible with straight vegetable oil, without the need for after-market modifications.

Walmart's truck fleet logs millions of miles each year, and the company planned to double the fleet's efficiency between 2005 and 2015. This truck is one of 15 based at Walmart's Buckeye, Arizona distribution center that was converted to run on a biofuel made from reclaimed cooking grease produced during food preparation at Walmart stores.

Vegetable oil can also be used in many older diesel engines that do not use common rail or unit injection electronic diesel injection systems. Due to the design of the combustion chambers in indirect injection engines, these are the best engines for use with vegetable oil. This system allows the relatively larger oil molecules more time to burn. Some older engines, especially Mercedes, are driven experimentally by enthusiasts without any conversion, a handful of drivers have experienced limited success with earlier pre-"Pumpe Duse" VW TDI engines and other similar engines with direct injection. Several companies, such as Elsbett or Wolf, have developed professional conversion kits and successfully installed hundreds of them over the last decades.

Oils and fats can be hydrogenated to give a diesel substitute. The resulting product is a straight-chain hydrocarbon with a high cetane number, low in aromatics and sulfur and does not contain oxygen. Hydrogenated oils can be blended with diesel in all proportions. They have several advantages over biodiesel, including good performance at low temperatures, no storage stability problems and no susceptibility to microbial attack.

Bioethers

Bioethers (also referred to as fuel ethers or oxygenated fuels) are cost-effective compounds that act as octane rating enhancers."Bioethers are produced by the reaction of reactive iso-olefins, such as iso-butylene, with bioethanol." Bioethers are created by wheat or sugar beet. They also enhance engine performance, whilst significantly reducing engine wear and toxic exhaust emissions. Though bioethers are likely to replace petroethers in the UK, it is highly unlikely they will become a fuel in and of itself due to the low energy density. Greatly reducing the amount of ground-level ozone emissions, they contribute to air quality.

When it comes to transportation fuel there are six ether additives- 1. Dimethyl Ether (DME) 2. Diethyl Ether (DEE) 3. Methyl Teritiary-Butyl Ether (MTBE) 4. Ethyl *ter*-butyl ether (ETBE) 5. T*er*-amyl methyl ether (TAME) 6. *Ter*-amyl ethyl Ether (TAEE)

The European Fuel Oxygenates Association (aka EFOA) credits Methyl Tertiary-Butyl Ether (MTBE) and Ethyl ter-butyl ether (ETBE) as the most commonly used ethers in fuel to replace lead. Ethers were brought into fuels in Europe in the 1970s to replace the highly toxic compound. Although Europeans still use Bio-ether additives, the US no longer has an oxygenate requirement therefore bio-ethers are no longer used as the main fuel additive.

Biogas

Biogas is methane produced by the process of anaerobic digestion of organic material by anaerobes. It can be produced either from biodegradable waste materials or by the use of energy crops fed into anaerobic digesters to supplement gas yields. The solid byproduct, digestate, can be used as a biofuel or a fertilizer.

- Biogas can be recovered from mechanical biological treatment waste processing systems.

  Note: Landfill gas, a less clean form of biogas, is produced in landfills through naturally occurring anaerobic digestion. If it escapes into the atmosphere, it is a potential greenhouse gas.

- Farmers can produce biogas from manure from their cattle by using anaerobic digesters.

Pipes carrying biogas

Syngas

Syngas, a mixture of carbon monoxide, hydrogen and other hydrocarbons, is produced by partial combustion of biomass, that is, combustion with an amount of oxygen that is not sufficient to convert the biomass completely to carbon dioxide and water. Before partial combustion, the biomass is dried, and sometimes pyrolysed. The resulting gas mixture, syngas, is more efficient than direct combustion of the original biofuel; more of the energy contained in the fuel is extracted.

- Syngas may be burned directly in internal combustion engines, turbines or high-temperature fuel cells. The wood gas generator, a wood-fueled gasification reactor, can be connected to an internal combustion engine.

- Syngas can be used to produce methanol, DME and hydrogen, or converted via the Fischer-Tropsch process to produce a diesel substitute, or a mixture of alcohols that can be blended into gasoline. Gasification normally relies on temperatures greater than 700 °C.

- Lower-temperature gasification is desirable when co-producing biochar, but results in syngas polluted with tar.

Solid Biofuels

Examples include wood, sawdust, grass trimmings, domestic refuse, charcoal, agricultural waste, nonfood energy crops, and dried manure.

When raw biomass is already in a suitable form (such as firewood), it can burn directly in a stove or furnace to provide heat or raise steam. When raw biomass is in an inconvenient form (such as sawdust, wood chips, grass, urban waste wood, agricultural residues), the typical process is to densify the biomass. This process includes grinding the raw biomass to an appropriate particulate size (known as hogfuel), which, depending on the densification type, can be from 1 to 3 cm (0.4 to 1.2 in), which is then concentrated into a fuel product. The current processes produce wood pellets, cubes, or pucks. The pellet process is most common in Europe, and is typically a pure wood product. The other types of densification are larger in size compared to a pellet, and are compatible with a broad range of input feedstocks. The resulting densified fuel is easier to transport and feed into thermal generation systems, such as boilers.

Industry has used sawdust, bark and chips for fuel for decades, primary in the pulp and paper industry, and also bagasse (spent sugar cane) fueled boilers in the sugar cane industry. Boilers in the range of 500,000 lb/hr of steam, and larger, are in routine operation, using grate, spreader stoker, suspension burning and fluid bed combustion. Utilities generate power, typically in the range of 5 to 50 MW, using locally available fuel. Other industries have also installed wood waste fueled boilers and dryers in areas with low cost fuel.

One of the advantages of biomass fuel is that it is often a byproduct, residue or waste-product of other processes, such as farming, animal husbandry and forestry. In theory, this means fuel and food production do not compete for resources, although this is not always the case.

A problem with the combustion of raw biomass is that it emits considerable amounts of pollutants, such as particulates and polycyclic aromatic hydrocarbons. Even modern pellet boilers generate much more pollutants than oil or natural gas boilers. Pellets made from agricultural residues are usually worse than wood pellets, producing much larger emissions of dioxins and chlorophenols.

In spite of the above noted study, numerous studies have shown biomass fuels have significantly less impact on the environment than fossil based fuels. Of note is the US Department of Energy Laboratory, operated by Midwest Research Institute Biomass Power and Conventional Fossil Systems with and without CO2 Sequestration – Comparing the Energy Balance, Greenhouse Gas Emissions and Economics Study. Power generation emits significant amounts of greenhouse gases (GHGs), mainly carbon dioxide (CO_2). Sequestering CO_2 from the power plant flue gas can significantly reduce the GHGs from the power plant itself, but this is not the total picture. CO_2 capture and sequestration consumes additional energy, thus lowering the plant's fuel-to-electricity efficiency. To compensate for this, more fossil fuel must be procured and consumed to make up for lost capacity.

Taking this into consideration, the global warming potential (GWP), which is a combination of CO_2, methane (CH_4), and nitrous oxide (N_2O) emissions, and energy balance of the system need to be examined using a life cycle assessment. This takes into account the upstream processes which remain constant after CO_2 sequestration, as well as the steps required for additional power generation. Firing biomass instead of coal led to a 148% reduction in GWP.

A derivative of solid biofuel is biochar, which is produced by biomass pyrolysis. Biochar made from agricultural waste can substitute for wood charcoal. As wood stock becomes scarce, this alternative is gaining ground. In eastern Democratic Republic of Congo, for example, biomass briquettes are being marketed as an alternative to charcoal to protect Virunga National Park from deforestation associated with charcoal production.

Second-generation (Advanced) Biofuels

Second generation biofuels, also known as advanced biofuels, are fuels that can be manufactured from various types of biomass. Biomass is a wide-ranging term meaning any source of organic carbon that is renewed rapidly as part of the carbon cycle. Biomass is derived from plant materials but can also include animal materials.

First generation biofuels are made from the sugars and vegetable oils found in arable crops, which can be easily extracted using conventional technology. In comparison, second generation biofuels are made from lignocellulosic biomass or woody crops, agricultural residues or waste, which makes it harder to extract the required fuel. A series of physical and chemical treatments might be required to convert lignocellulosic biomass to liquid fuels suitable for transportation.

Sustainable Biofuels

Biofuels in the form of liquid fuels derived from plant materials, are entering the market, driven mainly by the perception that they reduce climate gas emissions, and also by factors such as oil price spikes and the need for increased energy security. However, many of the biofuels that are currently being supplied have been criticised for their adverse impacts on the natural environment, food security, and land use. In 2008, the Nobel-prize winning chemist Paul J. Crutzen published findings that the release of nitrous oxide (N_2O) emissions in the production of biofuels means that overall they contribute more to global warming than the fossil fuels they replace.

The challenge is to support biofuel development, including the development of new cellulosic technologies, with responsible policies and economic instruments to help ensure that biofuel commercialization is sustainable. Responsible commercialization of biofuels represents an opportunity to enhance sustainable economic prospects in Africa, Latin America and Asia.

According to the Rocky Mountain Institute, sound biofuel production practices would not hamper food and fibre production, nor cause water or environmental problems, and would enhance soil fertility. The selection of land on which to grow the feedstocks is a critical component of the ability of biofuels to deliver sustainable solutions. A key consideration is the minimisation of biofuel competition for prime cropland.

Biofuels by Region

There are international organizations such as IEA Bioenergy, established in 1978 by the OECD International Energy Agency (IEA), with the aim of improving cooperation and information exchange between countries that have national programs in bioenergy research, development and deployment. The UN International Biofuels Forum is formed by Brazil, China, India, Pakistan, South Africa, the United States and the European Commission. The world leaders in biofuel development and use are Brazil, the United States, France, Sweden and Germany. Russia also has 22% of world's forest, and is a big biomass (solid biofuels) supplier. In 2010, Russian pulp and paper maker, Vyborgskaya Cellulose, said they would be producing pellets that can be used in heat and electricity generation from its plant in Vyborg by the end of the year. The plant will eventually produce about 900,000 tons of pellets per year, making it the largest in the world once operational.

Biofuels currently make up 3.1% of the total road transport fuel in the UK or 1,440 million litres. By 2020, 10% of the energy used in UK road and rail transport must come from renewable sources – this is the equivalent of replacing 4.3 million tonnes of fossil oil each year. Conventional biofuels are likely to produce between 3.7 and 6.6% of the energy needed in road and rail transport, while advanced biofuels could meet up to 4.3% of the UK's renewable transport fuel target by 2020.

Air Pollution

Biofuels are different from fossil fuels in regard to greenhouse gases but are similar to fossil fuels in that biofuels contribute to air pollution. Burning produces airborne carbon particulates, carbon monoxide and nitrous oxides. The WHO estimates 3.7 million premature deaths worldwide in 2012 due to air pollution. Brazil burns significant amounts of ethanol biofuel. Gas chromatograph studies were performed of ambient air in São Paulo, Brazil, and compared to Osaka, Japan, which does not burn ethanol fuel. Atmospheric Formaldehyde was 160% higher in Brazil, and Acetaldehyde was 260% higher.

Debates Regarding the Production and Use of Biofuel

There are various social, economic, environmental and technical issues with biofuel production and use, which have been discussed in the popular media and scientific journals. These include: the effect of moderating oil prices, the "food vs fuel" debate, food prices, poverty reduction potential, energy ratio, energy requirements, carbon emissions levels, sustainable biofuel production, deforestation and soil erosion, loss of biodiversity, impact on water resources, the possible modifications necessary to run the engine on biofuel, as well as energy balance and efficiency. The International Resource Panel, which provides independent scientific assessments and expert advice on a variety of resource-related themes, assessed the issues relating to biofuel use in its first report *Towards sustainable production and use of resources: Assessing Biofuels*. "Assessing Biofuels" outlined the wider and interrelated factors that need to be considered when deciding on the relative merits of pursuing one biofuel over another. It concluded that not all biofuels perform equally in terms of their impact on climate, energy security and ecosystems, and suggested that environmental and social impacts need to be assessed throughout the entire life-cycle.

Another issue with biofuel use and production is the US has changed mandates many times because the production has been taking longer than expected. The Renewable Fuel Standard (RFS) set by congress for 2010 was pushed back to at best 2012 to produce 100 million gallons of pure ethanol (not blended with a fossil fuel).

Current Research

Research is ongoing into finding more suitable biofuel crops and improving the oil yields of these crops. Using the current yields, vast amounts of land and fresh water would be needed to produce enough oil to completely replace fossil fuel usage. It would require twice the land area of the US to be devoted to soybean production, or two-thirds to be devoted to rapeseed production, to meet current US heating and transportation needs.

Specially bred mustard varieties can produce reasonably high oil yields and are very useful in crop rotation with cereals, and have the added benefit that the meal left over after the oil has been pressed out can act as an effective and biodegradable pesticide.

The NFESC, with Santa Barbara-based Biodiesel Industries, is working to develop biofuels technologies for the US navy and military, one of the largest diesel fuel users in the world. A group of Spanish developers working for a company called Ecofasa announced a new biofuel made from trash. The fuel is created from general urban waste which is treated by bacteria to produce fatty acids, which can be used to make biofuels.

Ethanol Biofuels

As the primary source of biofuels in North America, many organizations are conducting research in the area of ethanol production. The National Corn-to-Ethanol Research Center (NCERC) is a research division of Southern Illinois University Edwardsville dedicated solely to ethanol-based biofuel research projects. On the federal level, the USDA conducts a large amount of research regarding ethanol production in the United States. Much of this research is targeted toward the effect of ethanol production on domestic food markets. A division of the U.S. Department of Energy, the National Renewable Energy Laboratory (NREL), has also conducted various ethanol research projects, mainly in the area of cellulosic ethanol.

Cellulosic ethanol commercialization is the process of building an industry out of methods of turning cellulose-containing organic matter into fuel. Companies, such as Iogen, POET, and Abengoa, are building refineries that can process biomass and turn it into bioethanol. Companies, such as Diversa, Novozymes, and Dyadic, are producing enzymes that could enable a cellulosic ethanol future. The shift from food crop feedstocks to waste residues and native grasses offers significant opportunities for a range of players, from farmers to biotechnology firms, and from project developers to investors.

As of 2013, the first commercial-scale plants to produce cellulosic biofuels have begun operating. Multiple pathways for the conversion of different biofuel feedstocks are being used. In the next few years, the cost data of these technologies operating at commercial scale, and their relative performance, will become available. Lessons learnt will lower the costs of the industrial processes involved.

In parts of Asia and Africa where drylands prevail, sweet sorghum is being investigated as a potential source of food, feed and fuel combined. The crop is particularly suitable for growing in arid conditions, as it only extracts one seventh of the water used by sugarcane. In India, and other places, sweet sorghum stalks are used to produce biofuel by squeezing the juice and then fermenting into ethanol.

A study by researchers at the International Crops Research Institute for the Semi-Arid Tropics (ICRISAT) found that growing sweet sorghum instead of grain sorghum could increase farmers incomes by US$40 per hectare per crop because it can provide fuel in addition to food and animal feed. With grain sorghum currently grown on over 11 million hectares (ha) in Asia and on 23.4 million ha in Africa, a switch to sweet sorghum could have a considerable economic impact.

Algae Biofuels

From 1978 to 1996, the US NREL experimented with using algae as a biofuels source in the "Aquatic Species Program". A self-published article by Michael Briggs, at the UNH Biofuels Group, offers estimates for the realistic replacement of all vehicular fuel with biofuels by using algae that have a natural oil content greater than 50%, which Briggs suggests can be grown on algae ponds at wastewater treatment plants. This oil-rich algae can then be extracted from the system and processed into biofuels, with the dried remainder further reprocessed to create ethanol. The production of algae to harvest oil for biofuels has not yet been undertaken on a commercial scale, but feasibility studies have been conducted to arrive at the above yield estimate. In addition to its projected high yield, algaculture — unlike crop-based biofuels — does not entail a decrease in food production, since it requires neither farmland nor fresh water. Many companies are pursuing algae bioreactors for various purposes, including scaling up biofuels production to commercial levels. Prof. Rodrigo E. Teixeira from the University of Alabama in Huntsville demonstrated the extraction of biofuels lipids from wet algae using a simple and economical reaction in ionic liquids.

Jatropha

Several groups in various sectors are conducting research on *Jatropha curcas*, a poisonous shrub-like tree that produces seeds considered by many to be a viable source of biofuels feedstock oil. Much of this research focuses on improving the overall per acre oil yield of Jatropha through advancements in genetics, soil science, and horticultural practices.

SG Biofuels, a San Diego-based jatropha developer, has used molecular breeding and biotechnology to produce elite hybrid seeds that show significant yield improvements over first-generation varieties. SG Biofuels also claims additional benefits have arisen from such strains, including improved flowering synchronicity, higher resistance to pests and diseases, and increased cold-weather tolerance.

Plant Research International, a department of the Wageningen University and Research Centre in the Netherlands, maintains an ongoing Jatropha Evaluation Project that examines the feasibility of large-scale jatropha cultivation through field and laboratory experiments. The Center for Sustainable Energy Farming (CfSEF) is a Los Angeles-based nonprofit research organization dedicated to jatropha research in the areas of plant science, agronomy, and horticulture. Successful exploration of these disciplines is projected to increase jatropha farm production yields by 200-300% in the next 10 years.

Fungi

A group at the Russian Academy of Sciences in Moscow, in a 2008 paper, stated they had isolated large amounts of lipids from single-celled fungi and turned it into biofuels in an economically efficient manner. More research on this fungal species, *Cunninghamella japonica*, and others, is likely to appear in the near future. The recent discovery of a variant of the fungus *Gliocladium roseum* (later renamed Ascocoryne sarcoides) points toward the production of so-called myco-diesel from cellulose. This organism was recently discovered in the rainforests of northern Patagonia, and has the unique capability of converting cellulose into medium-length hydrocarbons typically found in diesel fuel. Many other fungi that can degrade cellulose and other polymers have been observed to produce molecules that are currently being engineered using organisms from other kingdoms, suggesting that fungi may play a large role in the bio-production of fuels in the future (reviewed in).

Animal Gut Bacteria

Microbial gastrointestinal flora in a variety of animals have shown potential for the production of biofuels. Recent research has shown that TU-103, a strain of *Clostridium* bacteria found in Zebra feces, can convert nearly any form of cellulose into butanol fuel. Microbes in panda waste are being investigated for their use in creating biofuels from bamboo and other plant materials. There has also been substantial research into the technology of using the gut microbiomes of wood-feeding insects for the conversion of lignocellulotic material into biofuel.

Greenhouse Gas Emissions

Some scientists have expressed concerns about land-use change in response to greater demand for crops to use for biofuel and the subsequent carbon emissions. The payback period, that is, the time it will take biofuels to pay back the carbon debt they acquire due to land-use change, has been estimated to be between 100 and 1000 years, depending on the specific instance and location of land-use change. However, no-till practices combined with cover-crop practices can reduce the payback period to three years for grassland conversion and 14 years for forest conversion.

A study conducted in the Tocantis State, in northern Brazil, found that many families were cutting down forests in order to produce two conglomerates of oilseed plants, the J. curcas (JC group) and the R. communis (RC group). This region is composed of 15% Amazonian rainforest with high biodiversity, and 80% Cerrado forest with lower biodiversity. During the study, the farmers that planted the JC group released over 2193 Mg CO_2, while losing 53-105 Mg CO_2 sequestration from deforestation; and the RC group farmers released 562 Mg CO_2, while losing 48-90 Mg CO_2 to be sequestered from forest depletion. The production of these types of biofuels not only led into an increased emission of carbon dioxide, but also to lower efficiency of forests to absorb the gases that these farms were emitting. This has to do with the amount of fossil fuel the production of fuel crops involves. In addition, the intensive use of monocropping agriculture requires large amounts of water irrigation, as well as of fertilizers, herbicides and pesticides. This does not only lead to poisonous chemicals to disperse on water runoff, but also to the emission of nitrous oxide (NO_2) as a fertilizer byproduct, which is three hundred times more efficient in producing a greenhouse effect than carbon dioxide (CO_2).

Converting rainforests, peatlands, savannas, or grasslands to produce food crop–based biofuels in Brazil, Southeast Asia, and the United States creates a "biofuel carbon debt" by releasing 17 to 420 times more CO_2 than the annual greenhouse gas (GHG) reductions that these biofuels would provide by displacing fossil fuels. Biofuels made from waste biomass or from biomass grown on abandoned agricultural lands incur little to no carbon debt.

Water use

In addition to water required to grow crops, biofuel facilities require significant process water.

Major Biofuels

Cellulosic Ethanol

Cellulosic ethanol is ethanol (ethyl alcohol) produced from cellulose (the stringy fiber of a plant) rather than from the plant's seeds or fruit. It is a biofuel produced from grasses, wood, algae, or other plants. The fibrous parts of the plants are mostly inedible to animals, including humans, except for ruminants (grazing, cud-chewing animals such as cows or sheep).

Considerable interest in cellulosic ethanol exists because it has the potential for strong economic importance. Growth of cellulose by plants is a mechanism that captures and stores solar energy chemically in nontoxic ways with resultant supplies that are easy to transport and store. Additionally, transport may be unneeded anyway, because grasses or trees can grow almost anywhere temperate. This is why commercially practical cellulosic ethanol is widely viewed as a next level of development for the biofuel industry that could reduce demand for oil and gas drilling and even nuclear power in ways that grain-based ethanol fuel alone cannot. Potential exists for the many benefits of carbonaceous liquid fuels and petrochemicals (which today's standard of living depends on) but in a carbon cycle–balanced and renewable way (recycling surface and atmosphere carbon instead of pumping underground carbon up into it and thus adding to it). Commercially practical cellulosic alcohol could also avoid one of the problems with today's conventional (grain-based) biofuels, which is that they set up competition for grain with food purposes, potentially driving up the price of food. To date, what stands in the way of these goals is that production of cellulosic alcohol is not yet sufficiently practical on a commercial scale.

Overview

Cellulosic ethanol is a type of biofuel produced from lignocellulose, a structural material that comprises much of the mass of plants. Lignocellulose is composed mainly of cellulose, hemicellulose and lignin. Corn stover, *Panicum virgatum* (switchgrass), *Miscanthus* grass species, wood chips and the byproducts of lawn and tree maintenance are some of the more popular cellulosic materials for ethanol production. Production of ethanol from lignocellulose has the advantage of abundant and diverse raw material compared to sources such as corn and cane sugars, but requires a greater amount of processing to make the sugar monomers available to the microorganisms typically used to produce ethanol by fermentation.

Switchgrass and *Miscanthus* are the major biomass materials being studied today, due to their high productivity per acre. Cellulose, however, is contained in nearly every natural, free-growing plant, tree, and bush, in meadows, forests, and fields all over the world without agricultural effort or cost needed to make it grow.

One of the benefits of cellulosic ethanol is it reduces greenhouse gas emissions (GHG) by 85% over reformulated gasoline. By contrast, starch ethanol (e.g., from corn), which most frequently uses natural gas to provide energy for the process, may not reduce GHG emissions at all depending on how the starch-based feedstock is produced. According to the National Academy of Sciences in 2011, there is no commercially viable bio-refinery in existence to convert lignocellulosic biomass to fuel. Absence of production of cellulosic ethanol in the quantities required by the regulation was the basis of a United States Court of Appeals for the District of Columbia decision announced January 25, 2013, voiding a requirement imposed on car and truck fuel producers in the United States by the Environmental Protection Agency requiring addition of cellulosic biofuels to their products. These issues, along with many other difficult production challenges, led George Washington University policy researchers to state that "in the short term, [cellulosic] ethanol cannot meet the energy security and environmental goals of a gasoline alternative."

History

The French chemist, Henri Braconnot, was the first to discover that cellulose could be hydrolyzed into sugars by treatment with sulfuric acid in 1819. The hydrolyzed sugar could then be processed to form ethanol through fermentation. The first commercialized ethanol production began in Germany in 1898, where acid was used to hydrolyze cellulose. In the United States, the Standard Alcohol Company opened the first cellulosic ethanol production plant in South Carolina in 1910. Later, a second plant was opened in Louisiana. However, both plants were closed after WWI due to economic reasons.

The first attempt at commercializing a process for ethanol from wood was done in Germany in 1898. It involved the use of dilute acid to hydrolyze the cellulose to glucose, and was able to produce 7.6 liters of ethanol per 100 kg of wood waste (18 US gal (68 L) per ton). The Germans soon developed an industrial process optimized for yields of around 50 US gallons (190 L) per ton of biomass. This process soon found its way to the US, culminating in two commercial plants operating in the southeast during WWI. These plants used what was called "the American Process" — a one-stage dilute sulfuric acid hydrolysis. Though the yields were half that of the original German process (25 US gallons (95 L) of ethanol per ton versus 50), the throughput of the American process was much higher. A drop in lumber production forced the plants to close shortly after the end of WWI. In the meantime, a small but steady amount of research on dilute acid hydrolysis continued at the USFS's Forest Products Laboratory. During World War II, the US again turned to cellulosic ethanol, this time for conversion to butadiene to produce synthetic rubber. The Vulcan Copper and Supply Company was contracted to construct and operate a plant to convert sawdust into ethanol. The plant was based on modifications to the original German Scholler process as developed by the Forest Products Laboratory. This plant achieved an ethanol yield of 50 US gal (190 L) per dry ton, but was still not profitable and was closed after the war.

With the rapid development of enzyme technologies in the last two decades, the acid hydrolysis process has gradually been replaced by enzymatic hydrolysis. Chemical pretreatment of the feedstock is required to prehydrolyze (separate) hemicellulose, so it can be more effectively converted into sugars. The dilute acid pretreatment is developed based on the early work on acid hydrolysis of wood at the USFS's Forest Products Laboratory. Recently, the Forest Products Laboratory together with the University of Wisconsin–Madison developed a sulfite pretreatment to overcome the recalcitrance of lignocellulose for robust enzymatic hydrolysis of wood cellulose.

US President George W. Bush, in his State of the Union address delivered January 31, 2006, proposed to expand the use of cellulosic ethanol. In his State of the Union Address on January 23, 2007, President Bush announced a proposed mandate for 35 billion US gallons (130,000,000 m^3) of ethanol by 2017. It is widely recognized that the maximum production of ethanol from corn starch is 15 billion US gallons (57,000,000 m^3) per year, implying a proposed mandate for production of some 20 billion US gallons (76,000,000 m^3) more per year of cellulosic ethanol by 2017. Bush's proposed plan includes $2 billion funding (from 2007 to 2017?) for cellulosic ethanol plants, with an additional $1.6 billion (from 2007 to 2017?) announced by the USDA on January 27, 2007.

In March 2007, the US government awarded $385 million in grants aimed at jump-starting ethanol production from nontraditional sources like wood chips, switchgrass, and citrus peels. Half of the six projects chosen will use thermochemical methods and half will use cellulosic ethanol methods.

The American company Range Fuels announced in July 2007 that it was awarded a construction permit from the state of Georgia to build the first commercial-scale 100-million-US-gallon (380,000 m^3)-per-year cellulosic ethanol plant in the US. Construction began in November, 2007. The Range Fuels plant was built in Soperton, GA, but was shut down in January 2011, without ever having produced any ethanol. It had received a $76 million grant from the US Department of Energy, plus $6 million from the State of Georgia, plus an $80 million loan guaranteed by the U.S. Biorefinery Assistance Program.

Production Methods

Bioreactor for cellulosic ethanol research.

The two ways of producing ethanol from cellulose are:

- Cellulolysis processes which consist of hydrolysis on pretreated lignocellulosic materials, using enzymes to break complex cellulose into simple sugars such as glucose, followed by fermentation and distillation.
- Gasification that transforms the lignocellulosic raw material into gaseous carbon monoxide and hydrogen. These gases can be converted to ethanol by fermentation or chemical catalysis.

As is normal for pure ethanol production, these methods include distillation.

Cellulolysis (Biological Approach)

The stages to produce ethanol using a biological approach are:

1. A "pretreatment" phase, to make the lignocellulosic material such as wood or straw amenable to hydrolysis
2. Cellulose hydrolysis (that is, cellulolysis) with cellulases, to break down the molecules into sugars
3. Separation of the sugar solution from the residual materials, notably lignin
4. Microbial fermentation of the sugar solution
5. Distillation to produce roughly 95% pure alcohol
6. Dehydration by molecular sieves to bring the ethanol concentration to over 99.5%

In 2010, a genetically engineered yeast strain was developed to produce its own cellulose-digesting enzymes. Assuming this technology can be scaled to industrial levels, it would eliminate one or more steps of cellulolysis, reducing both the time required and costs of production.

Pretreatment

Although lignocellulose is the most abundant plant material resource, its usability is curtailed by its rigid structure. As a result, an effective pretreatment is needed to liberate the cellulose from the lignin seal and its crystalline structure so as to render it accessible for a subsequent hydrolysis step. By far, most pretreatments are done through physical or chemical means. To achieve higher efficiency, both physical and chemical pretreatments are required. Physical pretreatment is often called size reduction to reduce biomass physical size. Chemical pretreatment is to remove chemical barriers so the enzymes can have access to cellulose for microbial reactions.

To date, the available pretreatment techniques include acid hydrolysis, steam explosion, ammonia fiber expansion, organosolv, sulfite pretreatment, AVAP® (SO2-ethanol-water) fractionation, alkaline wet oxidation and ozone pretreatment. Besides effective cellulose liberation, an ideal pretreatment has to minimize the formation of degradation products because of their inhibitory effects on subsequent hydrolysis and fermentation processes. The presence of inhibitors will not only further complicate the ethanol production but also increase the cost of production due to entailed detoxification steps. Even though pretreatment by acid hydrolysis is probably the oldest and most studied pretreatment technique, it produces several potent inhibitors including furfural and hydroxymethyl furfural (HMF) which are by far regarded as the most toxic inhibitors present in lignocellulosic hydrolysate. Ammonia Fiber Expansion (AFEX) is a promising pretreatment with no inhibitory effect in resulting hydrolysate.

Most pretreatment processes are not effective when applied to feedstocks with high lignin content, such as forest biomass. Organosolv, SPORL ('sulfite pretreatment to overcome recalcitrance of lignocellulose') and SO2-ethanol-water (AVAP®) processes are the three processes that can achieve over 90% cellulose conversion for forest biomass, especially those of softwood species. SPORL is the most energy efficient (sugar production per unit energy consumption in pretreatment) and robust process for pretreatment of forest biomass with very low production of fermentation inhibitors. Organosolv pulping is particularly effective for hardwoods and offers easy recovery of a hydrophobic lignin product by dilution and precipitation. AVAP® process effectively fractionates all types of lignocellulosics into clean highly digestible cellulose, undegraded hemicellulose sugars, reactive lignin and lignosulfonates, and is characterized by efficient recovery of chemicals.

Cellulolytic Processes

The cellulose molecules are composed of long chains of sugar molecules. In the hydrolysis of cellulose (that is, cellulolysis), these chains are broken down to free the sugar before it is fermented for alcohol production.

There are two major cellulose hydrolysis (cellulolysis) processes: a chemical reaction using acids, or an enzymatic reaction use cellulases.

Chemical Hydrolysis

In the traditional methods developed in the 19th century and at the beginning of the 20th century, hydrolysis is performed by attacking the cellulose with an acid. Dilute acid may be used under high heat and high pressure, or more concentrated acid can be used at lower temperatures and atmospheric pressure. A decrystalized cellulosic mixture of acid and sugars reacts in the presence of water to complete individual sugar molecules (hydrolysis). The product from this hydrolysis is then neutralized and yeast fermentation is used to produce ethanol. As mentioned, a significant obstacle to the dilute acid process is that the hydrolysis is so harsh that toxic degradation products are produced that can interfere with fermentation. BlueFire Renewables uses concentrated acid because it does not produce nearly as many fermentation inhibitors, but must be separated from the sugar stream for recycle [simulated moving bed (SMB) chromatographic separation, for example] to be commercially attractive.

Agricultural Research Service scientists found they can access and ferment almost all of the remaining sugars in wheat straw. The sugars are located in the plant's cell walls, which are notoriously difficult to break down. To access these sugars, scientists pretreated the wheat straw with alkaline peroxide, and then used specialized enzymes to break down the cell walls. This method produced 93 US gallons (350 L) of ethanol per ton of wheat straw.

Enzymatic Hydrolysis

Cellulose chains can be broken into glucose molecules by cellulase enzymes.

This reaction occurs at body temperature in the stomachs of ruminants such as cattle and sheep, where the enzymes are produced by microbes. This process uses several enzymes at various stages of this conversion. Using a similar enzymatic system, lignocellulosic materials can be enzymatically hydrolyzed at a relatively mild condition (50 °C and pH 5), thus enabling effective cellulose breakdown without the formation of byproducts that would otherwise inhibit enzyme activity. All major pretreatment methods, including dilute acid, require an enzymatic hydrolysis step to achieve high sugar yield for ethanol fermentation. Currently, most pretreatment studies have been laboratory-based, but companies are exploring means to transition from the laboratory to pilot, or production scale.

Various enzyme companies have also contributed significant technological breakthroughs in cellulosic ethanol through the mass production of enzymes for hydrolysis at competitive prices.

The fungus *Trichoderma reesei* is used by Iogen Corporation to secrete "specially engineered enzymes" for an enzymatic hydrolysis process. Their raw material (wood or straw) has to be pre-treated to make it amenable to hydrolysis.

Another Canadian company, SunOpta, uses steam explosion pretreatment, providing its technology to Verenium (formerly Celunol Corporation)'s facility in Jennings, Louisiana, Abengoa's facility in Salamanca, Spain, and a China Resources Alcohol Corporation in Zhaodong. The CRAC production facility uses corn stover as raw material.

Genencor and Novozymes have received United States Department of Energy funding for research into reducing the cost of cellulases, key enzymes in the production of cellulosic ethanol by enzymatic hydrolysis. A recent breakthrough in this regard was the discovery and inclusion of *lytic polysaccharide monooxygenases*. These enzymes are capable of boosting significantly the action of other cellulases by oxidatively attacking a polysaccharide substrate.

Other enzyme companies, such as Dyadic International, are developing genetically engineered fungi which would produce large volumes of cellulase, xylanase and hemicellulase enzymes, which can be used to convert agricultural residues such as corn stover, distiller grains, wheat straw and sugarcane bagasse and energy crops such as switchgrass into fermentable sugars which may be used to produce cellulosic ethanol.

In 2010, BP Biofuels bought out the cellulosic ethanol venture share of Verenium, which had itself been formed by the merger of Diversa and Celunol, and with which it jointly owned and operated a 1.4-million-US-gallon (5,300 m^3) per year demonstration plant in Jennings, LA, and the laboratory facilities and staff in San Diego, CA. BP Biofuels continues to operate these facilities, and has begun first phases to construct commercial facilities. Ethanol produced in the Jennings facility was shipped to London and blended with gasoline to provide fuel for the Olympics.

KL Energy Corporation, formerly KL Process Design Group, began commercial operation of a 1.5-million-US-gallon (5,700 m^3) per year cellulosic ethanol facility in Upton, WY in the last quarter of 2007. The Western Biomass Energy facility is currently achieving yields of 40–45 US gallons (150–170 L) per dry ton. It is the first operating commercial cellulosic ethanol facility in the nation. The KL Energy process uses a thermomechanical breakdown and enzymatic conversion process. The primary feedstock is soft wood, but lab tests have already proven the KL Energy process on wine pomace, sugarcane bagasse, municipal solid waste, and switchgrass.

Microbial Fermentation

Traditionally, baker's yeast (*Saccharomyces cerevisiae*), has long been used in the brewery industry to produce ethanol from hexoses (six-carbon sugars). Due to the complex nature of the carbohydrates present in lignocellulosic biomass, a significant amount of xylose and arabinose (five-carbon sugars derived from the hemicellulose portion of the lignocellulose) is also present in the hydrolysate. For example, in the hydrolysate of corn stover, approximately 30% of the total fermentable sugars is xylose. As a result, the ability of the fermenting microorganisms to use the whole range of sugars available from the hydrolysate is vital to increase the economic competitiveness of cellulosic ethanol and potentially biobased proteins.

In recent years, metabolic engineering for microorganisms used in fuel ethanol production has shown significant progress. Besides *Saccharomyces cerevisiae*, microorganisms such as *Zymomonas mobilis* and *Escherichia coli* have been targeted through metabolic engineering for cellulosic ethanol production.

Recently, engineered yeasts have been described efficiently fermenting xylose, and arabinose, and even both together. Yeast cells are especially attractive for cellulosic ethanol processes because they have been used in biotechnology for hundreds of years, are tolerant to high ethanol and inhibitor concentrations and can grow at low pH values to reduce bacterial contamination.

Combined Hydrolysis and Fermentation

Some species of bacteria have been found capable of direct conversion of a cellulose substrate into ethanol. One example is *Clostridium thermocellum*, which uses a complex cellulosome to break down cellulose and synthesize ethanol. However, *C. thermocellum* also produces other products during cellulose metabolism, including acetate and lactate, in addition to ethanol, lowering the efficiency of the process. Some research efforts are directed to optimizing ethanol production by genetically engineering bacteria that focus on the ethanol-producing pathway.

Gasification Process (Thermochemical Approach)

The gasification process does not rely on chemical decomposition of the cellulose chain (cellulolysis). Instead of breaking the cellulose into sugar molecules, the carbon in the raw material is converted into synthesis gas, using what amounts to partial combustion. The carbon monoxide, carbon dioxide and hydrogen may then be fed into a special kind of fermenter. Instead of sugar fermentation with yeast, this process uses *Clostridium ljungdahlii* bacteria. This microorganism will ingest carbon monoxide, carbon dioxide and hydrogen and produce ethanol and water. The process can thus be broken into three steps:

Fluidized Bed Gasifier in Güssing Burgenland Austria

1. Gasification — Complex carbon-based molecules are broken apart to access the carbon as carbon monoxide, carbon dioxide and hydrogen

2. Fermentation — Convert the carbon monoxide, carbon dioxide and hydrogen into ethanol using the *Clostridium ljungdahlii* organism

3. Distillation — Ethanol is separated from water

A recent study has found another *Clostridium* bacterium that seems to be twice as efficient in making ethanol from carbon monoxide as the one mentioned above.

Alternatively, the synthesis gas from gasification may be fed to a catalytic reactor where it is used to produce ethanol and other higher alcohols through a thermochemical process. This process can also generate other types of liquid fuels, an alternative concept successfully demonstrated by the Montreal-based company Enerkem at their facility in Westbury, Quebec.

Hemicellulose to Ethanol

Studies are intensively conducted to develop economic methods to convert both cellulose and hemicellulose to ethanol. Fermentation of glucose, the main product of cellulose hydrolyzate, to ethanol is an already established and efficient technique. However, conversion of xylose, the pentose sugar of hemicellulose hydrolyzate, is a limiting factor, especially in the presence of glucose. Moreover, it cannot be disregarded as hemicellulose will increase the efficiency and cost-effectiveness of cellulosic ethanol production.

Sakamoto (2012) et al. show the potential of genetic engineering microbes to express hemicellulase enzymes. The researchers created a recombinant Saccharomyces cerevisiae strain that was able to:

1. Hydrolyze hemicellulase through codisplaying endoxylanase on its cell surface,
2. Assimilate xylose by expression of xylose reductase and xylitol dehydrogenase.

The strain was able to convert rice straw hydrolyzate to ethanol, which contains hemicellulosic components. Moreover, it was able to produce 2.5x more ethanol than the control strain, showing the highly effective process of cell surface-engineering to produce ethanol.

Economics

The shift to a renewable fuel resource has been a target for many years now. However, most of its production is with the use of corn ethanol. In the year 2000, there was only 6.2 billion liters produced in the United States and it has expanded over 800% to 50 billion litres in just a decade (2010). Government pressures to shift to renewable fuel resources has been apparent since the U.S Environmental Protection Agency has implemented the 2007 Renewable Fuel Standard (RFS) to use a percentage of renewable fuel in products or face penalties. The shift to cellulosic ethanol production instead of corn has been strongly promoted by the US government. Even with these policies in place and the government attempting to create a market for cellulose ethanol, there was no commercial production of this fuel in 2010 and 2011. The Energy Independence and Security Act originally set goals of 100 million, 250 million and 500 million gallons for the years 2010, 2011 and 2012 respectively. However, as of 2012 it was projected that the production of cellulosic ethanol would be approximately 10.5 million far from its target. In 2007 alone, the US government provided 1 billion US dollars for cellulosic ethanol projects, while China invested 500 million US dollars into cellulosic ethanol research.

Due to the lack of existing commercialized plant data, it is difficult to determine the exact method of production that will be most commonly employed. Model systems try to compare different technologies costs, however these models cannot be applied to commercial-plant costs. Currently, there are many pilot and demonstration facilities open that show cellulosic production on a smaller scale. These main facilities are summarized in the table below.

Start-up costs for pilot scale lignocellulosic ethanol plants are high. On 28 February 2007, the U.S. Dept. of Energy announced $385 million in grant funding to six cellulosic ethanol plants. This grant funding accounts for 40% of the investment costs. The remaining 60% comes from the promoters of those facilities. Hence, a total of $1 billion will be invested for approximately 140-million-US-gallon (530,000 m^3) capacity. This translates into $7/annual gallon production capacity in capital investment costs for pilot plants; future capital costs are expected to be lower. Corn-to-ethanol plants cost roughly $1–3/annual gallon capacity, though the cost of the corn itself is considerably greater than for switchgrass or waste biomass.

As of 2007, ethanol is produced mostly from sugars or starches, obtained from fruits and grains. In contrast, cellulosic ethanol is obtained from cellulose, the main component of wood, straw, and much of the structure of plants. Since cellulose cannot be digested by humans, the production of cellulose does not compete with the production of food, other than conversion of land from food production to cellulose production (which has recently started to become an issue, due to rising wheat prices.) The price per ton of the raw material is thus much cheaper than that of grains or fruits. Moreover, since cellulose is the main component of plants, the whole plant can be harvested. This results in much better yields—up to 10 short tons per acre (22 t/ha), instead of 4-5 short tons/acre (9–11 t/ha) for the best crops of grain.

The raw material is plentiful. An estimated 323 million tons of cellulose-containing raw materials which could be used to create ethanol are thrown away each year in US alone. This includes 36.8 million dry tons of urban wood wastes, 90.5 million dry tons of primary mill residues, 45 million dry tons of forest residues, and 150.7 million dry tons of corn stover and wheat straw. Transforming them into ethanol using efficient and cost-effective hemi(cellulase) enzymes or other processes might provide as much as 30% of the current fuel consumption in the United States. Moreover, even land marginal for agriculture could be planted with cellulose-producing crops, such as switchgrass, resulting in enough production to substitute for all the current oil imports into the United States.

Paper, cardboard, and packaging comprise a substantial part of the solid waste sent to landfills in the United States each day, 41.26% of all organic municipal solid waste (MSW) according to California Integrated Waste Management Board's city profiles. These city profiles account for accumulation of 612.3 short tons (555.5 t) daily per landfill where an average population density of 2,413 per square mile persists. All these, except gypsum board, contain cellulose, which is transformable into cellulosic ethanol. This may have additional environmental benefits because decomposition of these products produces methane, a potent greenhouse gas.

Reduction of the disposal of solid waste through cellulosic ethanol conversion would reduce solid waste disposal costs by local and state governments. It is estimated that each person in the US throws away 4.4 lb (2.0 kg) of trash each day, of which 37% contains waste paper, which is largely cellulose. That computes to 244 thousand tons per day of discarded waste paper that contains cellulose. The raw material to produce cellulosic ethanol is not only free, it has a negative cost—i.e., ethanol producers can get paid to take it away.

In June 2006, a U.S. Senate hearing was told the current cost of producing cellulosic ethanol is US $2.25 per US gallon (US $0.59/litre), primarily due to the current poor conversion efficiency. At that price, it would cost about $120 to substitute a barrel of oil (42 US gallons (160 L)), taking into account the lower energy content of ethanol. However, the Department of Energy is optimistic and has requested a doubling of research funding. The same Senate hearing was told the research target was to reduce the cost of production to US $1.07 per US gallon (US $0.28/litre) by 2012. "The production of cellulosic ethanol represents not only a step toward true energy diversity for the country, but a very cost-effective alternative to fossil fuels. It is advanced weaponry in the war on oil," said Vinod Khosla, managing partner of Khosla Ventures, who recently told a Reuters Global Biofuels Summit that he could see cellulosic fuel prices sinking to $1 per gallon within ten years.

In September 2010, a report by Bloomberg analyzed the European biomass infrastructure and future refinery development. Estimated prices for a litre of ethanol in August 2010 are EUR 0.51 for 1g and 0.71 for 2g. The report suggested Europe should copy the current US subsidies of up to $50 per dry tonne.

Recently on October 25, 2012, BP, one of the leaders in fuel products, announced the cancellation of their proposed $350 million commercial-scale plant. It was estimated that the plant would be producing 36 million gallons a year at its location in Highlands County of Florida. BP has still provided 500 million US dollars for biofuel research at the Energy Biosciences Institute. General Motors (GM) has also invested into cellulosic companies more specifically Mascoma and Coskata. There are many other companies in construction or heading towards it. Abengoa is building a 25 million-gallon per year plant in \ technology platform based on the fungus Myceliophthora thermophila to convert lignocellulose into fermentable sugars. Poet is also in midst of producing a 200 million dollar, 25-million-gallon per year in Emmetsburg, Iowa. Mascoma now partnered with Valero has declared their intention to build a 20 million gallon per year in Kinross, Michigan. China Alcohol Resource Corporation has developed a 6.4 million liter cellulosic ethanol plant under continuous operation.

Also, since 2013, the Brazilian company GranBio is working to become a producer of biofuels and biochemicals. The family-held company is commissioning a 82 million liters per year (22 MMgy) cellulosic ethanol plant (2G ethanol) in the state of Alagoas, Brazil, which will be the first industrial facility of the group. GranBio's second generation ethanol facility is integrated to a first generation ethanol plant operated by Grupo Carlos Lyra, uses process technology from Beta Renewables, enzymes from Novozymes and yeast from DSM. Breaking ground in January 2013, the plant is in final commissioning. According to GranBio Annual Financial Records, the total investment was 208 million US Dollars.

Enzyme-cost Barrier

Cellulases and hemicellulases used in the production of cellulosic ethanol are more expensive compared to their first generation counterparts. Enzymes required for maize grain ethanol production cost 2.64-5.28 US dollars per cubic meter of ethanol produced. Enzymes for cellulosic ethanol production are projected to cost 79.25 US dollars, meaning they are 20-40 times more expensive. The cost differences are attributed to quantity required. The cellulase family of enzymes have a one to two order smaller magnitude of efficiency. Therefore, it requires 40 to 100 times more of the enzyme to be present in its production. For each ton of biomass it requires 15-25 kilograms of enzyme. More recent estimates are lower, suggesting 1 kg of enzyme per dry tonne of biomass feedstock. There is also relatively high capital costs associated with the long incubation times for the vessel that perform enzymatic hydrolysis. Altogether, enzymes comprise a significant portion of 20-40% for cellulosic ethanol production. A recent paper estimates the range at 13-36% of cash costs, with a key factor being how the cellulase enzyme is produced. For cellulase produced offsite, enzyme production amounts to 36% of cash cost. For enzyme produced onsite in a separate plant, the fraction is 29%; for integrated enzyme production, the faction is 13%. One of the key benefits of integrated production is that biomass instead of glucose is the enzyme growth medium. Biomass costs less, and it makes the resulting cellulosic ethanol a 100% second-generation biofuel, i.e., it uses no 'food for fuel'.

Feedstocks

In general there are two types of feedstocks: forest (woody) Biomass and agricultural biomass. In the US, about 1.4 billion dry tons of biomass can be sustainably produced annually. About 370 million tons or 30% are forest biomass. Forest biomass has higher cellulose and lignin content and lower hemicellulose and ash content than agricultural biomass. Because of the difficulties and low ethanol yield in fermenting pretreatment hydrolysate, especially those with very high 5 carbon hemicellulose sugars such as xylose, forest biomass has significant advantages over agricultural biomass. Forest biomass also has high density which significantly reduces transportation cost. It can be harvested year around which eliminates long term storage. The close to zero ash content of forest biomass significantly reduces dead load in transportation and processing. To meet the needs for biodiversity, forest biomass will be an important biomass feedstock supply mix in the future biobased economy. However, forest biomass is much more recalcitrant than agricultural biomass. Recently, the USDA Forest Products Laboratory together with the University of Wisconsin–Madison developed efficient technologies that can overcome the strong recalcitrance of forest (woody) biomass including those of softwood species that have low xylan content. Short-rotation intensive culture or tree farming can offer an almost unlimited opportunity for forest biomass production.

Woodchips from slashes and tree tops and saw dust from saw mills, and waste paper pulp are common forest biomass feedstocks for cellulosic ethanol production.

The following are a few examples of agricultural biomass:

Switchgrass (*Panicum virgatum*) is a native tallgrass prairie grass. Known for its hardiness and rapid growth, this perennial grows during the warm months to heights of 2–6 feet. Switchgrass can be grown in most parts of the United States, including swamplands, plains, streams, and along the shores & *interstate highways*. It is *self-seeding* (no tractor for sowing, only for mowing), resistant to many diseases and pests, & can produce high yields with low applications of fertilizer and other chemicals. It is also tolerant to poor soils, flooding, & drought; improves soil quality and prevents erosion due its type of root system.

Switchgrass is an approved cover crop for land protected under the federal Conservation Reserve Program (CRP). CRP is a government program that pays producers a fee for not growing crops on land on which crops recently grew. This program reduces soil erosion, enhances water quality, and increases wildlife habitat. CRP land serves as a habitat for upland game, such as pheasants and ducks, and a number of insects. Switchgrass for biofuel production has been considered for use on Conservation Reserve Program (CRP) land, which could increase ecological sustainability and lower the cost of the CRP program. However, CRP rules would have to be modified to allow this economic use of the CRP land.

Miscanthus × giganteus is another viable feedstock for cellulosic ethanol production. This species of grass is native to Asia and is the sterile triploid hybrid of *Miscanthus sinensis* and *Miscanthus sacchariflorus*. It can grow up to 12 feet (3.7 m) tall with little water or fertilizer input. Miscanthus is similar to switchgrass with respect to cold and drought tolerance and water use efficiency. Miscanthus is commercially grown in the European Union as a combustible energy source.

Corn cobs and corn stover are the most popular agricultural biomass.

It has been suggested that Kudzu may become a valuable source of biomass.

Environmental Effects

The environmental impact from the production of fuels is an important factor in determining its feasibility as an alternative to fossil fuels. Over the long run, small differences in production cost, environmental ramifications, and energy output may have large effects. It has been found that cellulosic ethanol can produce a positive net energy output. The reduction in green house gas (GHG) emissions from corn ethanol and cellulosic ethanol compared with fossil fuels is drastic. Corn ethanol may reduce overall GHG emissions by about 13%, while that figure is around 88% or greater for cellulosic ethanol. As well, cellulosic ethanol can reduce carbon dioxide emissions to nearly zero.

Croplands

A major concern for the viability of current alternative fuels is the cropland needed to produce the required materials. For example, the production of corn for corn ethanol fuel competes with cropland that may be used for food growth and other feedstocks. The difference between this and cellulosic ethanol production is that cellulosic material is widely available and is derived from a large resource of things. Some crops used for cellulosic ethanol production include switchgrass, corn stover, and hybrid poplar. These crops are fast-growing and can be grown on many types of land which makes them more versatile. Cellulosic ethanol can also be made from wood residues (chips and sawdust), municipal solid waste such as trash or garbage, paper and sewage sludge, cereal straws and grasses. It is particularly the non-edible portions of plant material which are used to make cellulosic ethanol, which also minimizes the potential cost of using food products in production.

The effectiveness of growing crops for the purpose of biomass can vary tremendously depending on the geographical location of the plot. For example, factors such as precipitation and sunlight exposure may greatly effect the energy input required to maintain the crops, and therefore effect the overall energy output. A study done over five years showed that growing and managing switchgrass exclusively as a biomass energy crop can produce 500% or more renewable energy than is consumed during production. The levels of GHG emissions and carbon dioxide were also drastically decreased from using cellulosic ethanol compared with traditional gasoline.

Corn-based vs. Grass-based

Summary of Searchinger et al. comparison of corn ethanol and gasoline GHG emissions with and without land use change (Grams of CO_2 released per megajoule of energy in fuel)				
Fuel type (U.S.)	**Carbon intensity**	**Reduction GHG**	**Carbon intensity + ILUC**	**Reduction GHG**
Gasoline	**92**	-	**92**	-
Corn ethanol	74	-20%	177	+93%
Cellulosic ethanol	28	-70%	138	+50%
Notes: Calculated using default assumptions for 2015 scenario for ethanol in E85. Gasoline is a combination of conventional and reformulated gasoline.				

In 2008, there was only a small amount of switchgrass dedicated for ethanol production. In order for it to be grown on a large-scale production it must compete with existing uses of agricultural land, mainly for the production of crop commodities. Of the United States' 2.26 billion acres (9.1 million km^2) of unsubmerged land, 33% are forestland, 26% pastureland and grassland, and 20% crop land. A study done by the U.S. Departments of Energy and Agriculture in 2005 determined whether there were enough available land resources to sustain production of over 1 billion dry tons of biomass annually to replace 30% or more of the nation's current use of liquid transportation fuels. The study found that there could be 1.3 billion dry tons of biomass available for ethanol use, by making little changes in agricultural and forestry practices and meeting the demands for forestry products, food, and fiber. A recent study done by the University of Tennessee reported that as many as 100 million acres (400,000 km^2, or 154,000 sq mi) of cropland and pasture will need to be allocated to switchgrass production in order to offset petroleum use by 25 percent.

Currently, corn is easier and less expensive to process into ethanol in comparison to cellulosic ethanol. The Department of Energy estimates that it costs about $2.20 per gallon to produce cellulosic ethanol, which is twice as much as ethanol from corn. Enzymes that destroy plant cell wall tissue cost 30 to 50 cents per gallon of ethanol compared to 3 cents per gallon for corn. The Department of Energy hopes to reduce production cost to $1.07 per gallon by 2012 to be effective. However, cellulosic biomass is cheaper to produce than corn, because it requires fewer inputs, such as energy, fertilizer, herbicide, and is accompanied by less soil erosion and improved soil fertility. Additionally, nonfermentable and unconverted solids left after making ethanol can be burned to provide the fuel needed to operate the conversion plant and produce electricity. Energy used to run corn-based ethanol plants is derived from coal and natural gas. The Institute for Local Self-Reliance estimates the cost of cellulosic ethanol from the first generation of commercial plants will be in the $1.90–$2.25 per gallon range, excluding incentives. This compares to the current cost of $1.20–$1.50 per gallon for ethanol from corn and the current retail price of over $4.00 per gallon for regular gasoline (which is subsidized and taxed).

One of the major reasons for increasing the use of biofuels is to reduce greenhouse gas emissions. In comparison to gasoline, ethanol burns cleaner, thus putting less carbon dioxide and overall pollution in the air. Additionally, only low levels of smog are produced from combustion. According to the U.S. Department of Energy, ethanol from cellulose reduces greenhouse gas emission by 86 percent when compared to gasoline and to corn-based ethanol, which decreases emissions by 52 percent. Carbon dioxide gas emissions are shown to be 85% lower than those from gasoline. Cellulosic ethanol contributes little to the greenhouse effect and has a five times better net energy balance than corn-based ethanol. When used as a fuel, cellulosic ethanol releases less sulfur, carbon monoxide, particulates, and greenhouse gases. Cellulosic ethanol should earn producers carbon reduction credits, higher than those given to producers who grow corn for ethanol, which is about 3 to 20 cents per gallon.

It takes 0.76 J of energy from fossil fuels to produce 1 J worth of ethanol from corn. This total includes the use of fossil fuels used for fertilizer, tractor fuel, ethanol plant operation, etc. Research has shown that fossil fuel can produce over five times the volume of ethanol from prairie grasses, according to Terry Riley, President of Policy at the Theodore Roosevelt Conservation Partnership. The United States Department of Energy concludes that corn-based ethanol provides 26 percent more energy than it requires for production, while cellulosic ethanol provides 80 percent more energy. Cellulosic ethanol yields 80 percent more energy than is required to grow and convert it. The process of turning corn into ethanol requires about 1700 times (by volume) as much water as ethanol produced. Additionally, it leaves 12 times its volume in waste. Grain ethanol uses only the edible portion of the plant.

U.S. Environmental Protection Agency Draft life cycle GHG emissions reduction results for different time horizon and discount rate approaches (includes indirect land use change effects)		
Fuel Pathway	**100 years + 2% discount rate**	**30 years + 0% discount rate**
Corn ethanol (natural gas dry mill)	-16%	+5%
Corn ethanol (Best case NG DM)	-39%	-18%
Corn ethanol (coal dry mill)	+13%	+34%
Corn ethanol (biomass dry mill)	-39%	-18%
Corn ethanol (biomass dry mill with combined heat and power)	-47%	-26%
Brazilian sugarcane ethanol	-44%	-26%
Cellulosic ethanol from switchgrass	-128%	-124%
Cellulosic ethanol from corn stover	-115%	-116%
Notes: (1) Dry mill (DM) plants grind the entire kernel and generally produce only one primary co-product: distillers grains with solubles (DGS). (2) Best case plants produce wet distillers grains co-product.		

Cellulose is not used for food and can be grown in all parts of the world. The entire plant can be used when producing cellulosic ethanol. Switchgrass yields twice as much ethanol per acre than corn. Therefore, less land is needed for production and thus less habitat fragmentation. Biomass materials require fewer inputs, such as fertilizer, herbicides, and other chemicals that can pose risks to wildlife. Their extensive roots improve soil quality, reduce erosion, and increase nutrient capture. Herbaceous energy crops reduce soil erosion by greater than 90%, when compared to conventional commodity crop production. This can translate into improved water quality for rural communities. Additionally, herbaceous energy crops add organic material to depleted soils and can increase soil carbon, which can have a direct effect on climate change, as soil carbon can absorb carbon dioxide in the air. As compared to commodity crop production, biomass reduces surface runoff and nitrogen transport. Switchgrass provides an environment for diverse wildlife habitation, mainly insects and ground birds. Conservation Reserve Program (CRP) land is composed of perennial grasses, which are used for cellulosic ethanol, and may be available for use.

For years American farmers have practiced row cropping, with crops such as sorghum and corn. Because of this, much is known about the effect of these practices on wildlife. The most significant effect of increased corn ethanol would be the additional land that would have to be converted to agricultural use and the increased erosion and fertilizer use that goes along with agricultural production. Increasing our ethanol production through the use of corn could produce negative effects on wildlife, the magnitude of which will depend on the scale of production and whether the land used for this increased production was formerly idle, in a natural state, or planted with other row crops. Another consideration is whether to plant a switchgrass monoculture or use a variety of grasses and other vegetation. While a mixture of vegetation types likely would provide better wildlife habitat, the technology has not yet developed to allow the processing of a mixture of different grass species or vegetation types into bioethanol. Of course, cellulosic ethanol production is still in its infancy, and the possibility of using diverse vegetation stands instead of monocultures deserves further exploration as research continues.

A study by Nobel Prize winner Paul Crutzen found ethanol produced from corn had a "net climate warming" effect when compared to oil when the full life cycle assessment properly considers the nitrous oxide (N2o) emissions that occur during corn ethanol production. Crutzen found that crops with less nitrogen demand, such as grasses and woody coppice species, have more favourable climate impacts.

Cellulosic Ethanol Commercialization

Cellulosic ethanol commercialization is the process of building an industry out of methods of turning cellulose-containing organic matter into fuel. Companies such as Iogen, POET, and Abengoa are building refineries that can process biomass and turn it into ethanol, while companies such as DuPont, Diversa, Novozymes, and Dyadic are producing enzymes which could enable a cellulosic ethanol future. The shift from food crop feedstocks to waste residues and native grasses offers significant opportunities for a range of players, from farmers to biotechnology firms, and from project developers to investors.

The cellulosic ethanol industry developed some new commercial-scale plants in 2008. In the United States, plants totaling 12 million liters (3.17 million gal) per year were operational, and an additional 80 million liters (21.1 million gal.) per year of capacity - in 26 new plants - was under construction. In Canada, capacity of 6 million liters per year was operational. In Europe, several plants were operational in Germany, Spain, and Sweden, and capacity of 10 million liters per year was under construction.

Italy-based Mossi & Ghisolfi Group broke ground for its 13 MMgy cellulosic ethanol facility in northwestern Italy on April 12, 2011. The project will be the largest cellulosic ethanol project in the world, 10 times larger than any of the currently operating demonstration-scale facilities.

Commercial Cellulosic Ethanol Plants in the U.S. (Operational or Under Construction)		
Company	**Location**	**Feedstock**
Abengoa Bioenergy	Hugoton, KS	Wheat straw
BlueFire Ethanol	Irvine, CA	Multiple sources
Colusa Biomass Energy Corporation	Sacramento, CA	Waste rice straw
Coskata	Warrenville, IL	Biomass, Agricultural and Municipal wastes
DuPont	Vonore, TN	Corn cobs, switchgrass
DuPont	Nevada, IA	Corn stover
Fulcrum BioEnergy	Reno, NV	Municipal solid waste
Gulf Coast Energy	Mossy Head, FL	Wood waste
KL Energy Corp.	Upton, WY	Wood
Mascoma	Lansing, MI	Wood
POET-DSM Advanced Biofuels	Emmetsburg, IA	Corn cobs, husks, and stover
Range Fuels	Treutlen County, GA	Wood waste
SunOpta	Little Falls, MN	Wood chips
SweetWater Energy	Rochester, NY	Multiple Sources
US Envirofuels	Highlands County, FL	Sweet sorghum
Xethanol	Auburndale, FL	Citrus peels

Vegetable Oil Fuel

Vegetable oil can be used as an alternative fuel in diesel engines and in heating oil burners. When vegetable oil is used directly as a fuel, in either modified or unmodified equipment, it is referred to as straight vegetable oil (SVO) or pure plant oil (PPO). Conventional diesel engines can be modified to help ensure that the viscosity of the vegetable oil is low enough to allow proper atomization of the fuel. This prevents incomplete combustion, which would damage the engine by causing a build-up of carbon. Straight vegetable oil can also be blended with conventional diesel or processed into biodiesel or bioliquids for use under a wider range of conditions.

History

Rudolf Diesel was the father of the engine which bears his name. His first attempts were to design an engine to run on coal dust, but later designed his engine to run on vegetable oil. The idea, he hoped, would make his engines more attractive to farmers having a source of fuel readily available. In a 1912 presentation to the British Institute of Mechanical Engineers, he cited a number of efforts in this area and remarked, "The fact that fat oils from vegetable sources can be used may seem insignificant today, but such oils may perhaps become in course of time of the same importance as some natural mineral oils and the tar products are now."

Periodic petroleum shortages spurred research into vegetable oil as a diesel substitute during the 1930s and 1940s, and again in the 1970s and early 1980s when straight vegetable oil enjoyed its highest level of scientific interest. The 1970s also saw the formation of the first commercial enterprise to allow consumers to run straight vegetable oil in their automobiles, Elsbett of Germany. In the 1990s Bougainvillea conflict, islanders cut off from oil supplies due to a blockade used coconut oil to fuel their vehicles.

Application and Usability

Modified Fuel Systems

Most diesel car engines are suitable for the use of straight vegetable oil (SVO), also commonly called pure plant oil (PPO), with certain modifications. Principally, the viscosity and surface tension of the SVO/PPO must be reduced by preheating it, typically by using waste heat from the engine or electricity, otherwise poor atomization, incomplete combustion and carbonization may result. One common solution is to add a heat exchanger and an additional fuel tank for the petrodiesel or biodiesel blend and to switch between this additional tank and the main tank of SVO/PPO. The engine is started on diesel, switched over to vegetable oil as soon as it is warmed up and switched back to diesel shortly before being switched off to ensure that no vegetable oil remains in the engine or fuel lines when it is started from cold again. In colder climates it is often necessary to heat the vegetable oil fuel lines and tank as it can become very viscous and even solidify.

Single tank conversions have been developed, largely in Germany, which have been used throughout Europe. These conversions are designed to provide reliable operation with rapeseed oil that meets the German rapeseed oil fuel standard DIN 51605. Modifications to the engines cold start regime assist combustion on start up and during the engine warm up phase. Suitably modified indirect injection (IDI) engines have proven to be operable with 100% PPO down to temperatures of −10 °C (14 °F). Direct injection (DI) engines generally have to be preheated with a block heater or diesel fired heater. The exception is the VW Tdi (Turbocharged Direct Injection) engine for which a number of German companies offer single tank conversions. For long term durability it has been found necessary to increase the oil change frequency and to pay increased attention to engine maintenance.

Unmodified Indirect Injection Engines

Many cars powered by indirect injection engines supplied by in-line injection pumps, or mechanical Bosch injection pumps are capable of running on pure SVO/PPO in all but winter temperatures. Indirect injection Mercedes-Benz vehicles with in-line injection pumps and cars featuring the PSA XUD engine tend to perform reasonably, especially as the latter is normally equipped with a coolant heated fuel filter. Engine reliability would depend on the condition of the engine. Attention to maintenance of the engine, particularly of the fuel injectors, cooling system and glow plugs will help to provide longevity. Ideally the engine would be converted.

Vegetable Oil Blending

The relatively high kinematic viscosity of vegetable oils must be reduced to make them compatible with conventional compression-ignition engines and fuel systems. Cosolvent blending is a low-cost and easy-to-adapt technology that reduces viscosity by diluting the vegetable oil with a low-molecular-weight solvent. This blending, or "cutting", has been done with diesel fuel, kerosene, and gasoline, amongst others; however, opinions vary as to the efficacy of this. Noted problems include higher rates of wear and failure in fuel pumps and piston rings when using blends.

Home Heating

When liquid fuels made from biomass are used for energy purposes other than transport, they are called bioliquids.

With often minimal modification, most residential furnaces and boilers that are designed to burn No. 2 heating oil can be made to burn either biodiesel or filtered, preheated waste vegetable oil (WVO). If cleaned at home by the consumer, WVO can result in considerable savings. Many restaurants will receive a minimal amount for their used cooking oil, and processing to biodiesel is fairly simple and inexpensive. Burning filtered WVO directly is somewhat more problematic, since it is much more viscous; nonetheless, its burning can be accomplished with suitable preheating. WVO can thus be an economical heating option for those with the necessary mechanical and experimental aptitude.

Combined Heat and Power

A number of companies offer compressed ignition engine generators optimized to run on plant oils where the waste engine heat is recovered for heating.

Properties

The main form of SVO/PPO used in the UK is rapeseed oil (also known as canola oil, primarily in the United States and Canada) which has a freezing point of −10 °C (14 °F). However the use of sunflower oil, which gels at around −12 °C (10 °F), is currently being investigated as a means of improving cold weather starting. Unfortunately oils with lower gelling points tend to be less saturated (leading to a higher iodine number) and polymerize more easily in the presence of atmospheric oxygen.

Material Compatibility

Polymerization also has been consequentially linked to catastrophic component failures such as injection pump shaft seizure and breakage, injector tip failure leading to various and/or combustion chamber components damaged. Most metallurgical problems such as corrosion and electrolysis are related to water based contamination or poor choices of plumbing (such as copper or Zinc) which can cause gelling- even with petroleum based fuels.

Temperature Effects

Some Pacific island nations are using coconut oil as fuel to reduce their expenses and their dependence on imported fuels while helping stabilize the coconut oil market. Coconut oil is only usable where temperatures do not drop below 17 degrees Celsius (63 degrees Fahrenheit), unless two-tank SVO/PPO kits or other tank-heating accessories, etc. are used. Fortunately, the same techniques developed to use, for example, canola and other oils in cold climates can be implemented to make coconut oil usable in temperatures lower than 17 degrees Celsius (63 degrees Fahrenheit)

Availability

Recycled Vegetable Oil

Recycled vegetable oil, also termed used vegetable oil (UVO), waste vegetable oil (WVO), used cooking oil (UCO), or yellow grease (in commodities exchange), is recovered from businesses and industry that use the oil for cooking.

As of 2000, the United States was producing in excess of 11 billion liters (2.9 billion U.S. gallons) of recycled vegetable oil annually, mainly from industrial deep fryers in potato processing plants, snack food factories and fast food restaurants. If all those 11 billion liters could be recycled and used to replace the energy equivalent amount of petroleum (an ideal case), almost 1% of US oil consumption could be offset. Use of used vegetable oil as a direct fuel competes with some other uses of the commodity, which has effects on its price as a fuel and increases its cost as an input to the other uses as well.

Virgin Vegetable Oil

Virgin vegetable oil, also termed pure plant oil or straight vegetable oil, is extracted from plants solely for use as fuel. In contrast to used vegetable oil, is not a byproduct of other industries, and thus its prospects for use as fuel are not limited by the capacities of other industries. Production of vegetable oils for use as fuels is theoretically limited only by the agricultural capacity of a given economy. However, doing so detracts from the supply of other uses of pure vegetable oil.

Legal Implications

Taxation of Fuel

Taxation on SVO/PPO as a road fuel varies from country to country, and it is possible the revenue departments in many countries are even unaware of its use, or feel it too insignificant to legislate. Germany used to have 0% taxation, resulting in it being a leader in most developments of the fuel use. However SVO/PPO as a road fuel began to be taxed at 0,09 €/liter from 1 January 2008 in Germany, with incremental rises up to 0,45 €/liter by 2012. However, in Australia it has become illegal to produce any fuel if it is to be sold unless a license to do so is granted by the federal government. This is a chargeable offense with a fine of up to 20,000 dollars but this bracket may alter circumstantially. Also a jail term may result if offenders are aware of the illegality of selling the fuel.

Ethanol Fuel

The Saab 9-3 SportCombi BioPower was the second E85 flexifuel model introduced by Saab in the Swedish market.

Ethanol fuel is ethyl alcohol, the same type of alcohol found in alcoholic beverages, used as fuel. It is most often used as a motor fuel, mainly as a biofuel additive for gasoline. The first production car running entirely on ethanol was the Fiat 147, introduced in 1978 in Brazil by Fiat. Nowadays, cars are able to run using 100% ethanol fuel or a mix of ethanol and gasoline (aka flex-fuel). It is commonly made from biomass such as corn or sugarcane. World ethanol production for transport fuel tripled between 2000 and 2007 from 17 billion to more than 52 billion liters. From 2007 to 2008, the share of ethanol in global gasoline type fuel use increased from 3.7% to 5.4%. In 2011 worldwide ethanol fuel production reached 22.36 billion U.S. liquid gallons (bg) (84.6 billion liters), with the United States as the top producer with 13.9 bg (52.6 billion liters), accounting for 62.2% of global production, followed by Brazil with 5.6 bg (21.1 billion liters). Ethanol fuel has a "gasoline gallon equivalency" (GGE) value of 1.5 US gallons (5.7 L), which means 1.5 gallons of ethanol produces the energy of one gallon of gasoline.

Ethanol fuel is widely used in Brazil and in the United States, and together both countries were responsible for 87.1% of the world's ethanol fuel production in 2011. Most cars on the road today in the U.S. can run on blends of up to 10% ethanol, and ethanol represented 10% of the U.S. gasoline fuel supply derived from domestic sources in 2011. Since 1976 the Brazilian government has made it mandatory to blend ethanol with gasoline, and since 2007 the legal blend is around 25% ethanol and 75% gasoline (E25). By December 2011 Brazil had a fleet of 14.8 million flex-fuel automobiles and light trucks and 1.5 million flex-fuel motorcycles that regularly use neat ethanol fuel (known as E100).

Bioethanol is a form of quasi-renewable energy that can be produced from agricultural feedstocks. It can be made from very common crops such as hemp, sugarcane, potato, cassava and corn. There has been considerable debate about how useful bioethanol is in replacing gasoline. Concerns about its production and use relate to increased food prices due to the large amount of arable land required for crops, as well as the energy and pollution balance of the whole cycle of ethanol production, especially from corn. Recent developments with cellulosic ethanol production and commercialization may allay some of these concerns.

Cellulosic ethanol offers promise because cellulose fibers, a major and universal component in plant cells walls, can be used to produce ethanol. According to the International Energy Agency, cellulosic ethanol could allow ethanol fuels to play a much bigger role in the future.

Chemistry

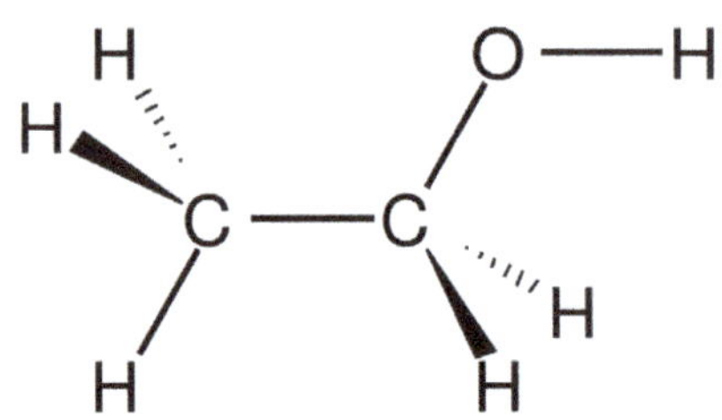

Structure of ethanol molecule. All bonds are single bonds

During ethanol fermentation, glucose and other sugars in the corn (or sugarcane or other crops) are converted into ethanol and carbon dioxide.

$$C_6H_{12}O_6 \rightarrow 2\ C_2H_5OH + 2\ CO_2 + \text{heat}$$

Ethanol fermentation is not 100% selective with other side products such as acetic acid, glycols and many other products produced. They are mostly removed during ethanol purification. Fermentation takes place in an aqueous solution. The resulting solution has an ethanol content of around 15%. Ethanol is subsequently isolated and purified by a combination of adsorption and distillation.

During combustion, ethanol reacts with oxygen to produce carbon dioxide, water, and heat:

$$C_2H_5OH + 3\ O_2 \rightarrow 2\ CO_2 + 3\ H_2O + \text{heat}$$

Starch and cellulose molecules are strings of glucose molecules. It is also possible to generate ethanol out of cellulosic materials. That, however, requires a pretreatment that splits the cellulose into glycose molecules and other sugars that subsequently can be fermented. The resulting product is called cellulosic ethanol, indicating its source.

Ethanol may also be produced industrially from ethylene by hydration of the double bond in the presence of catalysts and high temperature.

$$C_2H_4 + H_2O \rightarrow C_2H_5OH$$

By far the largest fraction of the global ethanol production, however, is produced by fermentation.

Sources

Sugar cane harvest

About 5% of the ethanol produced in the world in 2003 was actually a petroleum product. It is made by the catalytic hydration of ethylene with sulfuric acid as the catalyst. It can also be obtained via ethylene or acetylene, from calcium carbide, coal, oil gas, and other sources. Two million tons of petroleum-derived ethanol are produced annually. The principal suppliers are plants in the United States, Europe, and South Africa. Petroleum derived ethanol (synthetic ethanol) is chemically identical to bio-ethanol and can be differentiated only by radiocarbon dating.

Bio-ethanol is usually obtained from the conversion of carbon-based feedstock. Agricultural feedstocks are considered renewable because they get energy from the sun using photosynthesis, provided that all minerals required for growth (such as nitrogen and phosphorus) are returned to the land. Ethanol can be produced from a variety of feedstocks such as sugar cane, bagasse, miscanthus, sugar beet, sorghum, grain, switchgrass, barley, hemp, kenaf, potatoes, sweet potatoes, cassava, sunflower, fruit, molasses, corn, stover, grain, wheat, straw, cotton, other biomass, as well as many types of cellulose waste and harvesting, whichever has the best well-to-wheel assessment.

Switchgrass

An alternative process to produce bio-ethanol from algae is being developed by the company Algenol. Rather than grow algae and then harvest and ferment it, the algae grow in sunlight and produce ethanol directly, which is removed without killing the algae. It is claimed the process can produce 6,000 US gallons per acre (56,000 litres per ha) per year compared with 400 US gallons per acre (3,750 l/ha) for corn production.

Cornfield in South Africa

Currently, the first generation processes for the production of ethanol from corn use only a small part of the corn plant: the corn kernels are taken from the corn plant and only the starch, which represents about 50% of the dry kernel mass, is transformed into ethanol. Two types of second generation processes are under development. The first type uses enzymes and yeast fermentation to convert the plant cellulose into ethanol while the second type uses pyrolysis to convert the whole plant to either a liquid bio-oil or a syngas. Second generation processes can also be used with plants such as grasses, wood or agricultural waste material such as straw.

Production

The basic steps for large-scale production of ethanol are: microbial (yeast) fermentation of sugars, distillation, dehydration (requirements vary, see Ethanol fuel mixtures, below), and denaturing (optional). Prior to fermentation, some crops require saccharification or hydrolysis of carbohydrates such as cellulose and starch into sugars. Saccharification of cellulose is called cellulolysis. Enzymes are used to convert starch into sugar.

Fermentation

Ethanol is produced by microbial fermentation of the sugar. Microbial fermentation currently only works directly with sugars. Two major components of plants, starch and cellulose, are both made of sugars—and can, in principle, be converted to sugars for fermentation. Currently, only the sugar (e.g., sugar cane) and starch (e.g., corn) portions can be economically converted. There is much activity in the area of cellulosic ethanol, where the cellulose part of a plant is broken down to sugars and subsequently converted to ethanol.

Distillation

Ethanol plant in West Burlington, Iowa

For the ethanol to be usable as a fuel, the majority of the water must be removed. Most of the water is removed by distillation, but the purity is limited to 95–96% due to the formation of a low-boiling water-ethanol azeotrope with maximum (95.6% m/m (96.5% v/v) ethanol and 4.4% m/m (3.5% v/v) water). This mixture is called hydrous ethanol and can be used as a fuel alone, but unlike anhydrous ethanol, hydrous ethanol is not miscible in all ratios with gasoline, so the water fraction is typically removed in further treatment to burn in combination with gasoline in gasoline engines.

Ethanol plant in Sertãozinho, Brazil.

Dehydration

There are basically three dehydration processes to remove the water from an azeotropic ethanol/water mixture. The first process, used in many early fuel ethanol plants, is called azeotropic distillation and consists of adding benzene or cyclohexane to the mixture. When these components are added to the mixture, it forms a heterogeneous azeotropic mixture in vapor–liquid-liquid equilibrium, which when distilled produces anhydrous ethanol in the column bottom, and a vapor mixture of water, ethanol, and cyclohexane/benzene.

When condensed, this becomes a two-phase liquid mixture. The heavier phase, poor in the entrainer (benzene or cyclohexane), is stripped of the entrainer and recycled to the feed—while the lighter phase, with condensate from the stripping, is recycled to the second column. Another early method, called extractive distillation, consists of adding a ternary component that increases ethanol's relative volatility. When the ternary mixture is distilled, it produces anhydrous ethanol on the top stream of the column.

With increasing attention being paid to saving energy, many methods have been proposed that avoid distillation altogether for dehydration. Of these methods, a third method has emerged and has been adopted by the majority of modern ethanol plants. This new process uses molecular sieves to remove water from fuel ethanol. In this process, ethanol vapor under pressure passes through a bed of molecular sieve beads. The bead's pores are sized to allow absorption of water while excluding ethanol. After a period of time, the bed is regenerated under vacuum or in the flow of inert atmosphere (e.g. N_2) to remove the absorbed water. Two beds are often used so that one is available to absorb water while the other is being regenerated. This dehydration technology can account for energy saving of 3,000 btus/gallon (840 kJ/L) compared to earlier azeotropic distillation.

Post-production Water Issues

Ethanol is hygroscopic, meaning it absorbs water vapor directly from the atmosphere. Because absorbed water dilutes the fuel value of the ethanol and may cause phase separation of ethanol-gasoline blends (which causes engine stall), containers of ethanol fuels must be kept tightly sealed. This high miscibility with water means that ethanol cannot be efficiently shipped through modern pipelines, like liquid hydrocarbons, over long distances.

The fraction of water that an ethanol-gasoline fuel can contain without phase separation increases with the percentage of ethanol. This shows, for example, that E30 can have up to about 2% water. If there is more than about 71% ethanol, the remainder can be any proportion of water or gasoline and phase separation does not occur. The fuel mileage declines with increased water content. The increased solubility of water with higher ethanol content permits E30 and hydrated ethanol to be put in the same tank since any combination of them always results in a single phase. Somewhat less water is tolerated at lower temperatures. For E10 it is about 0.5% v/v at 70 F and decreases to about 0.23% v/v at −30 F.

Consumer Production Systems

While biodiesel production systems have been marketed to home and business users for many years, commercialized ethanol production systems designed for end-consumer use have lagged in the marketplace. In 2008, two different companies announced home-scale ethanol production systems. The AFS125 Advanced Fuel System from Allard Research and Development is capable of producing both ethanol and biodiesel in one machine, while the E-100 MicroFueler from E-Fuel Corporation is dedicated to ethanol only.

Engines

Fuel Economy

Ethanol contains approx. 34% less energy per unit volume than gasoline, and therefore in theory, burning pure ethanol in a vehicle reduces miles per US gallon 34%, given the same fuel economy, compared to burning pure gasoline. However, since ethanol has a higher octane rating, the engine can be made more efficient by raising its compression ratio. Using a variable turbocharger, the compression ratio can be optimized for the fuel, making fuel economy almost constant for any blend.

For E10 (10% ethanol and 90% gasoline), the effect is small (~3%) when compared to conventional gasoline, and even smaller (1–2%) when compared to oxygenated and reformulated blends. For E85 (85% ethanol), the effect becomes significant. E85 produces lower mileage than gasoline, and requires more frequent refueling. Actual performance may vary depending on the vehicle. Based on EPA tests for all 2006 E85 models, the average fuel economy for E85 vehicles resulted 25.56% lower than unleaded gasoline. The EPA-rated mileage of current United States flex-fuel vehicles should be considered when making price comparisons, but E85 is a high performance fuel, with an octane rating of about 94–96, and should be compared to premium.

Cold Start During the Winter

The Brazilian 2008 Honda Civic flex-fuel has outside direct access to the secondary reservoir gasoline tank in the front right side, the corresponding fuel filler door is shown by the arrow.

High ethanol blends present a problem to achieve enough vapor pressure for the fuel to evaporate and spark the ignition during cold weather (since ethanol tends to increase fuel enthalpy of vaporization). When vapor pressure is below 45 kPa starting a cold engine becomes difficult. To avoid this problem at temperatures below 11 °C (52 °F), and to reduce ethanol higher emissions during cold weather, both the US and the European markets adopted E85 as the maximum blend to be used in their flexible fuel vehicles, and they are optimized to run at such a blend. At places with harsh cold weather, the ethanol blend in the US has a seasonal reduction to E70 for these very cold regions, though it is still sold as E85. At places where temperatures fall below −12 °C (10 °F) during the winter, it is recommended to install an engine heater system, both for gasoline and E85 vehicles. Sweden has a similar seasonal reduction, but the ethanol content in the blend is reduced to E75 during the winter months.

Brazilian flex fuel vehicles can operate with ethanol mixtures up to E100, which is hydrous ethanol (with up to 4% water), which causes vapor pressure to drop faster as compared to E85 vehicles. As a result, Brazilian flex vehicles are built with a small secondary gasoline reservoir located near the engine. During a cold start pure gasoline is injected to avoid starting problems at low temperatures. This provision is particularly necessary for users of Brazil's southern and central regions, where temperatures normally drop below 15 °C (59 °F) during the winter. An improved flex engine generation was launched in 2009 that eliminates the need for the secondary gas storage tank. In March 2009 Volkswagen do Brasil launched the Polo E-Flex, the first Brazilian flex fuel model without an auxiliary tank for cold start.

Fuel Mixtures

In many countries cars are mandated to run on mixtures of ethanol. All Brazilian light-duty vehicles are built to operate for an ethanol blend of up to 25% (E25), and since 1993 a federal law requires mixtures between 22% and 25% ethanol, with 25% required as of mid July 2011. In the United States all light-duty vehicles are built to operate normally with an ethanol blend of 10% (E10). At the end of 2010 over 90 percent of all gasoline sold in the U.S. was blended with ethanol. In January 2011 the U.S. Environmental Protection Agency (EPA) issued a waiver to authorize up to 15% of ethanol blended with gasoline (E15) to be sold only for cars and light pickup trucks with a model year of 2001 or newer.

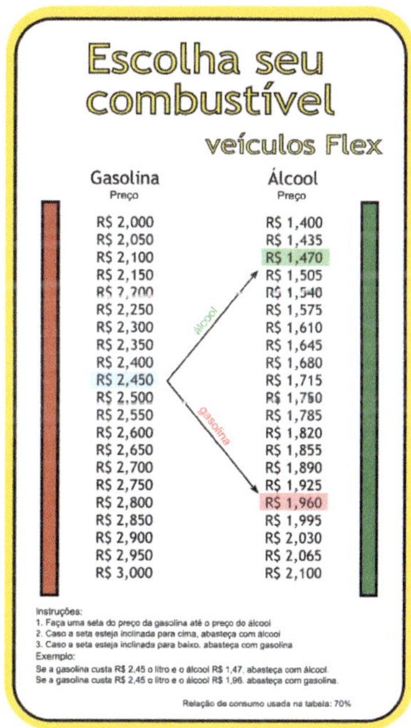

Hydrated ethanol × gasoline type C price table for use in Brazil

Beginning with the model year 1999, an increasing number of vehicles in the world are manufactured with engines that can run on any fuel from 0% ethanol up to 100% ethanol without modification. Many cars and light trucks (a class containing minivans, SUVs and pickup trucks) are designed to be flexible-fuel vehicles using ethanol blends up to 85% (E85) in North America and Europe, and up to 100% (E100) in Brazil. In older model years, their engine systems contained alcohol sensors in the fuel and/or oxygen sensors in the exhaust that provide input to the engine control computer to adjust the fuel injection to achieve stochiometric (no residual fuel or free oxygen in the exhaust) air-to-fuel ratio for any fuel mix. In newer models, the alcohol sensors have been removed, with the computer using only oxygen and airflow sensor feedback to estimate alcohol content. The engine control computer can also adjust (advance) the ignition timing to achieve a higher output without pre-ignition when it predicts that higher alcohol percentages are present in the fuel being burned. This method is backed up by advanced knock sensors – used in most high performance gasoline engines regardless of whether they are designed to use ethanol or not – that detect pre-ignition and detonation.

EPA's E15 label required to be displayed in all E15 fuel dispensers in the U.S.

Other Engine Configurations

ED95 Engines

Since 1989 there have also been ethanol engines based on the diesel principle operating in Sweden. They are used primarily in city buses, but also in distribution trucks and waste collectors. The engines, made by Scania, have a modified compression ratio, and the fuel (known as ED95) used is a mix of 93.6% ethanol and 3.6% ignition improver, and 2.8% denaturants. The ignition improver makes it possible for the fuel to ignite in the diesel combustion cycle. It is then also possible to use the energy efficiency of the diesel principle with ethanol. These engines have been used in the United Kingdom by Reading Transport but the use of bioethanol fuel is now being phased out.

Dual-fuel Direct-injection

A 2004 MIT study and an earlier paper published by the Society of Automotive Engineers identified a method to exploit the characteristics of fuel ethanol substantially more efficiently than mixing it with gasoline. The method presents the possibility of leveraging the use of alcohol to achieve definite improvement over the cost-effectiveness of hybrid electric. The improvement consists of using dual-fuel direct-injection of pure alcohol (or the azeotrope or E85) and gasoline, in any ratio up to 100% of either, in a turbocharged, high compression-ratio, small-displacement engine having performance similar to an engine having twice the displacement. Each fuel is carried separately, with a much smaller tank for alcohol. The high-compression (for higher efficiency) engine runs on ordinary gasoline under low-power cruise conditions. Alcohol is directly injected into the cylinders (and the gasoline injection simultaneously reduced) only when necessary to suppress 'knock' such as when significantly accelerating. Direct cylinder injection raises the already high octane rating of ethanol up to an effective 130. The calculated over-all reduction of gasoline use and CO_2 emission is 30%. The consumer cost payback time shows a 4:1 improvement over turbo-diesel and a 5:1 improvement over hybrid. The problems of water absorption into pre-mixed gasoline (causing phase separation), supply issues of multiple mix ratios and cold-weather starting are also avoided.

Increased Thermal Efficiency

In a 2008 study, complex engine controls and increased exhaust gas recirculation allowed a compression ratio of 19.5 with fuels ranging from neat ethanol to E50. Thermal efficiency up to approximately that for a diesel was achieved. This would result in the fuel economy of a neat ethanol vehicle to be about the same as one burning gasoline.

Fuel Cells Powered by an Ethanol Reformer

In June 2016, Nissan announced plans to develop fuel cell vehicles powered by ethanol rather than hydrogen, the fuel of choice by the other car manufacturers that have developed and commercialized fuel cell vehicles, such as the Hyundai Tucson FCEV, Toyota Mirai, and Honda FCX Clarity. The main advantage of this technical approach is that it would be cheaper and easier to deploy the fueling infrastructure than setting up the one required to deliver hydrogen at high pressures, as each hydrogen fueling station cost US$1 million to US$2 million to build.

Nissan plans to create a technology that uses liquid ethanol fuel as a source to generate hydrogen within the vehicle itself. The technology uses heat to reform ethanol into hydrogen to feed what is known as a solid oxide fuel cell (SOFC). The fuel cell generates electricity to supply power to the electric motor driving the wheels, through a battery that handles peak power demands and stores regenerated energy. The vehicle would include a tank for a blend of water and ethanol, which is fed into an onboard reformer that splits it into pure hydrogen and carbon dioxide. According to Nissan, the liquid fuel could be an ethanol-water blend at a 55:45 ratio. Nissan expects to commercialize its technology by 2020.

Experience by Country

The world's top ethanol fuel producers in 2011 were the United States with 13.9 billion U.S. liquid gallons (bg) (52.60 billion liters) and Brazil with 5.6 bg (21.1 billion liters), accounting together for 87.1% of world production of 22.36 billion US gallons (84.6 billion liters). Strong incentives, coupled with other industry development initiatives, are giving rise to fledgling ethanol industries in countries such as Germany, Spain, France, Sweden, China, Thailand, Canada, Colombia, India, Australia, and some Central American countries.

Annual fuel ethanol production by country (2007–2011) Top 10 countries/regional blocks (Millions of U.S. liquid gallons per year)						
World rank	Country/Region	2011	2010	2009	2008	2007
1	United States	13,900.00	13,231.00	10,938.00	9,235.00	6,485.00

2	Brazil	5,573.24	6,921.54	6,577.89	6,472.20	5,019.20
3	EU	1,199.31	1,176.88	1,039.52	733.60	570.30
4	China	554.76	541.55	541.55	501.90	486.00
5	Thailand			435.20	89.80	79.20
6	Canada	462.30	356.63	290.59	237.70	211.30
7	India			91.67	66.00	52.80
8	Colombia			83.21	79.30	74.90
9	Australia	87.20	66.04	56.80	26.40	26.40
10	Other			247.27		
	World Total	22,356.09	22,946.87	19,534.99	17,335.20	13,101.70

Environment

Energy Balance

Energy balance		
Country	**Type**	**Energy balance**
United States	Corn ethanol	1.3
Germany	Biodiesel	2.5
Brazil	Sugarcane ethanol	8
United States	Cellulosic ethanol†	2–36††

† experimental, not in commercial production

†† depending on production method

All biomass goes through at least some of these steps: it needs to be grown, collected, dried, fermented, distilled, and burned. All of these steps require resources and an infrastructure. The total amount of energy input into the process compared to the energy released by burning the resulting ethanol fuel is known as the energy balance (or "energy returned on energy invested"). Figures compiled in a 2007 by *National Geographic Magazine* point to modest results for corn ethanol produced in the US: one unit of fossil-fuel energy is required to create 1.3 energy units from the resulting ethanol. The energy balance for sugarcane ethanol produced in Brazil is more favorable, with one unit of fossil-fuel energy required to create 8 from the ethanol. Energy balance estimates are not easily produced, thus numerous such reports have been generated that are contradictory. For instance, a separate survey reports that production of ethanol from sugarcane, which requires a tropical climate to grow productively, returns from 8 to 9 units of energy for each unit expended, as compared to corn, which only returns about 1.34 units of fuel energy for each unit of energy expended. A 2006 University of California Berkeley study, after analyzing six separate studies, concluded that producing ethanol from corn uses much less petroleum than producing gasoline.

Carbon dioxide, a greenhouse gas, is emitted during fermentation and combustion. This is canceled out by the greater uptake of carbon dioxide by the plants as they grow to produce the biomass. When compared to gasoline, depending on the production method, ethanol releases less greenhouse gases.

Air Pollution

Compared with conventional unleaded gasoline, ethanol is a particulate-free burning fuel source that combusts with oxygen to form carbon dioxide, carbon monoxide, water and aldehydes. The Clean Air Act requires the addition of oxygenates to reduce carbon monoxide emissions in the United States. The additive MTBE is currently being phased out due to ground water contamination, hence ethanol becomes an attractive alternative additive. Current production methods include air pollution from the manufacturer of macronutrient fertilizers such as ammonia.

A study by atmospheric scientists at Stanford University found that E85 fuel would increase the risk of air pollution deaths relative to gasoline by 9% in Los Angeles, US: a very large, urban, car-based metropolis that is a worst-case scenario. Ozone levels are significantly increased, thereby increasing photochemical smog and aggravating medical problems such as asthma.

Brazil burns significant amounts of ethanol biofuel. Gas chromatograph studies were performed of ambient air in São Paulo, Brazil, and compared to Osaka, Japan, which does not burn ethanol fuel. Atmospheric Formaldehyde was 160% higher in Brazil, and Acetaldehyde was 260% higher.

Carbon Dioxide

The calculation of exactly how much carbon dioxide is produced in the manufacture of bioethanol is a complex and inexact process, and is highly dependent on the method by which the ethanol is produced and the assumptions made in the calculation. A calculation should include:

- The cost of growing the feedstock
- The cost of transporting the feedstock to the factory
- The cost of processing the feedstock into bioethanol

Such a calculation may or may not consider the following effects:

- The cost of the change in land use of the area where the fuel feedstock is grown.

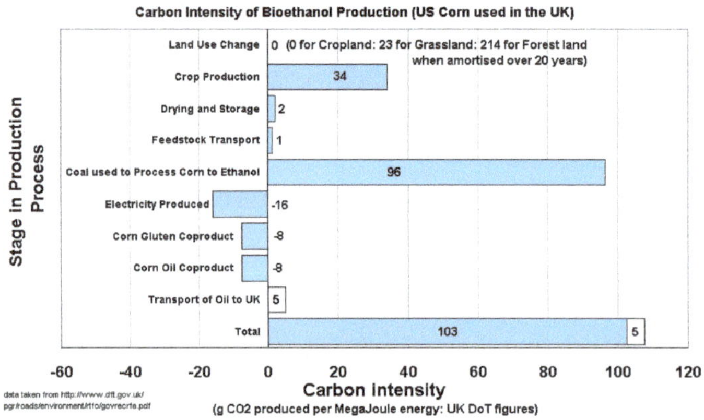

UK government calculation of carbon intensity of corn bioethanol grown in the US and burnt in the UK.

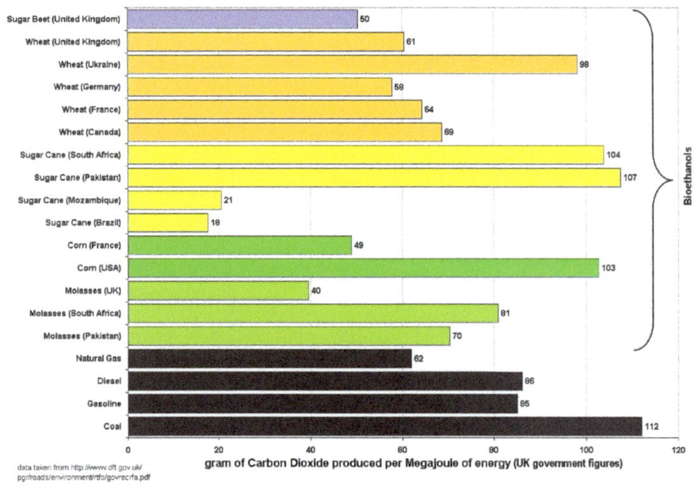

Graph of UK figures for the carbon intensity of bioethanol and fossil fuels. This graph assumes that all bioethanols are burnt in their country of origin and that previously existing cropland is used to grow the feedstock.

- The cost of transportation of the bioethanol from the factory to its point of use
- The efficiency of the bioethanol compared with standard gasoline
- The amount of carbon dioxide produced at the tail pipe.
- The benefits due to the production of useful bi-products, such as cattle feed or electricity.

The graph on the right shows figures calculated by the UK government for the purposes of the Renewable transport fuel obligation.

The January 2006 Science article from UC Berkeley's ERG, estimated reduction from corn ethanol in GHG to be 13% after reviewing a large number of studies. In a correction to that article released shortly after publication, they reduce the estimated value to

7.4%. A National Geographic Magazine overview article (2007) puts the figures at 22% less CO_2 emissions in production and use for corn ethanol compared to gasoline and a 56% reduction for cane ethanol. Carmaker Ford reports a 70% reduction in CO_2 emissions with bioethanol compared to petrol for one of their flexible-fuel vehicles.

An additional complication is that production requires tilling new soil which produces a one-off release of GHG that it can take decades or centuries of production reductions in GHG emissions to equalize. As an example, converting grass lands to corn production for ethanol takes about a century of annual savings to make up for the GHG released from the initial tilling.

Change in Land Use

Large-scale farming is necessary to produce agricultural alcohol and this requires substantial amounts of cultivated land. University of Minnesota researchers report that if all corn grown in the U.S. were used to make ethanol it would displace 12% of current U.S. gasoline consumption. There are claims that land for ethanol production is acquired through deforestation, while others have observed that areas currently supporting forests are usually not suitable for growing crops. In any case, farming may involve a decline in soil fertility due to reduction of organic matter, a decrease in water availability and quality, an increase in the use of pesticides and fertilizers, and potential dislocation of local communities. New technology enables farmers and processors to increasingly produce the same output using less inputs.

Cellulosic ethanol production is a new approach that may alleviate land use and related concerns. Cellulosic ethanol can be produced from any plant material, potentially doubling yields, in an effort to minimize conflict between food needs vs. fuel needs. Instead of utilizing only the starch by-products from grinding wheat and other crops, cellulosic ethanol production maximizes the use of all plant materials, including gluten. This approach would have a smaller carbon footprint because the amount of energy-intensive fertilisers and fungicides remain the same for higher output of usable material. The technology for producing cellulosic ethanol is currently in the commercialization stage.

Using Biomass for Electricity Instead of Ethanol

Converting biomass to electricity for charging electric vehicles may be a more "climate-friendly" transportation option than using biomass to produce ethanol fuel, according to an analysis published in Science in May 2009 Researchers continue to search for more cost-effective developments in both cellulosic ethanol and advanced vehicle batteries.

Health Costs of Ethanol Emissions

For each billion ethanol-equivalent gallons of fuel produced and combusted in the US, the combined climate-change and health costs are \$469 million for gasoline, \$472–952 million for corn ethanol depending on biorefinery heat source (natural gas, corn stover, or coal) and technology, but only $123–208 million for cellulosic ethanol depending on feedstock (prairie biomass, Miscanthus, corn stover, or switchgrass).

Efficiency of Common Crops

As ethanol yields improve or different feedstocks are introduced, ethanol production may become more economically feasible in the US. Currently, research on improving ethanol yields from each unit of corn is underway using biotechnology. Also, as long as oil prices remain high, the economical use of other feedstocks, such as cellulose, become viable. By-products such as straw or wood chips can be converted to ethanol. Fast growing species like switchgrass can be grown on land not suitable for other cash crops and yield high levels of ethanol per unit area.

Crop	Annual yield (Liters/hectare, US gal/acre)	Green-house-gas savings vs. petrol[a]	Comments
Sugar cane	6800–8000 L/ha, 727–870 g/acre	87%–96%	Long-season annual grass. Used as feedstock for most bioethanol produced in Brazil. Newer processing plants burn residues not used for ethanol to generate electricity. Grows only in tropical and subtropical climates.
Miscanthus	7300 L/ha, 780 g/acre	37%–73%	Low-input perennial grass. Ethanol production depends on development of cellulosic technology.
Switchgrass	3100–7600 L/ha, 330–810 g/acre	37%–73%	Low-input perennial grass. Ethanol production depends on development of cellulosic technology. Breeding efforts underway to increase yields. Higher biomass production possible with mixed species of perennial grasses.
Poplar	3700–6000 L/ha, 400–640 g/acre	51%–100%	Fast-growing tree. Ethanol production depends on development of cellulosic technology. Completion of genomic sequencing project will aid breeding efforts to increase yields.
Sweet sorghum	2500–7000 L/ha, 270–750 g/acre	No data	Low-input annual grass. Ethanol production possible using existing technology. Grows in tropical and temperate climates, but highest ethanol yield estimates assume multiple crops per year (possible only in tropical climates). Does not store well.
Corn	3100–4000 L/ha, 330–424 g/acre	10%–20%	High-input annual grass. Used as feedstock for most bioethanol produced in USA. Only kernels can be processed using available technology; development of commercial cellulosic technology would allow stover to be used and increase ethanol yield by 1,100 – 2,000 litres/ha.

Source (except those indicated): Nature 444 (7 December 2006): 673–676.
[a] – Savings of GHG emissions assuming no land use change (using existing crop lands).

Reduced Petroleum Imports and Costs

One rationale given for extensive ethanol production in the U.S. is its benefit to energy security, by shifting the need for some foreign-produced oil to domestically produced energy sources. Production of ethanol requires significant energy, but current U.S. production derives most of that energy from coal, natural gas and other sources, rather than oil. Because 66% of oil consumed in the U.S. is imported, compared to a net surplus of coal and just 16% of natural gas (figures from 2006), the displacement of oil-based fuels to ethanol produces a net shift from foreign to domestic U.S. energy sources.

According to a 2008 analysis by Iowa State University, the growth in US ethanol production has caused retail gasoline prices to be US $0.29 to US $0.40 per gallon lower than would otherwise have been the case.

Motorsport

Leon Duray qualified third for the 1927 Indianapolis 500 auto race with an ethanol-fueled car. The IndyCar Series adopted a 10% ethanol blend for the 2006 season, and a 98% blend in 2007.

In drag racing, there are Top Alcohol classes for dragsters and funny cars since the 1970s.

The American Le Mans Series sports car championship introduced E10 in the 2007 season to replace pure gasoline. In the 2008 season, E85 was allowed in the GT class and teams began switching to it.

In 2011, the three national NASCAR stock car series mandated a switch from gasoline to E15, a blend of Sunoco GTX unleaded racing fuel and 15% ethanol.

Australia's V8 Supercar championship uses United E85 for its racing fuel.

Stock Car Brasil Championship runs on neat ethanol, E100.

Ethanol fuel may also be utilized as a rocket fuel. As of 2010, small quantities of ethanol are used in lightweight rocket-racing aircraft.

Replacement Cooking Fuel

Project Gaia is a U.S. non-governmental, non-profit organization involved in the creation of a commercially viable household market for alcohol-based fuels in Ethiopia and other countries in the developing world. The project considers alcohol fuels to be a solution to fuel shortages, environmental damage, and public health issues caused by traditional cooking in the developing world. Targeting poor and marginalized communities that face health issues from cooking over polluting fires, Gaia currently works in Ethiopia, Nigeria, Brazil, Haiti, and Madagascar, and is in the planning stage of projects in several other countries.

Research

Ethanol plant in Turner County, South Dakota

Ethanol research focuses on alternative sources, novel catalysts and production processes.

In 2013, INEOS began initial operation of a bio-ethanol plant from vegetative material and wood waste.

The bacterium E.coli when genetically engineered with cow rumen genes and enzymes can produce ethanol from corn stover.

An alternative technology allows for the production of biodiesel from grain that has already been used to produce ethanol.

Another approach uses feed stocks such as municipal waste, recycled products, rice hulls, sugarcane bagasse, wood chips or switchgrass.

Butanol Fuel

Butanol may be used as a fuel in an internal combustion engine. Because its longer hydrocarbon chain causes it to be fairly non-polar, it is more similar to gasoline than it is to ethanol. Butanol has been demonstrated to work in vehicles designed for use with gasoline without modification. It has a four link hydrocarbon chain. It can be produced from biomass (as "biobutanol") as well as fossil fuels (as "petrobutanol"), but biobutanol and petrobutanol have the same chemical properties.

Production of Biobutanol

Butanol from biomass is called biobutanol. It can be used in unmodified gasoline engines.

Technologies

Biobutanol can be produced by fermentation of biomass by the A.B.E. process. The process uses the bacterium *Clostridium acetobutylicum*, also known as the *Weizmann organism*. It was Chaim Weizmann who first used this bacterium for the production of acetone from starch (with the main use of acetone being the making of Cordite) in 1916. The butanol was a by-product of this fermentation (twice as much butanol was produced). The process also creates a recoverable amount of H_2 and a number of other by-products: acetic, lactic and propionic acids, isopropanol and ethanol.

Biobutanol can also be made using *Ralstonia eutropha* H16. This process requires the use of an electro-bioreactor and the input of carbon dioxide and electricity.

The difference from ethanol production is primarily in the fermentation of the feedstock and minor changes in distillation. The feedstocks are the same as for ethanol: energy crops such as sugar beets, sugar cane, corn grain, wheat and cassava, prospective non-food energy crops such as switchgrass and even guayule in North America, as well as agricultural byproducts such as bagasse, straw and corn stalks. According to DuPont, existing bioethanol plants can cost-effectively be retrofitted to biobutanol production.

Additionally, butanol production from biomass and agricultural byproducts could be more efficient (i.e. unit engine motive power delivered per unit solar energy consumed) than ethanol or methanol production.

Algae Butanol

Biobutanol can be made entirely with solar energy and nutrients, from algae (called Solalgal Fuel) or diatoms. Current yield is low.

Research

Although biofuel demand has risen to over one billion liters (about 260 million US gallons) yearly, fermentation remains a largely inefficient method of butanol production. Under normal living conditions, *Clostridium* bacterial communities have a low yield of butanol per gram of glucose. Obtaining higher yields of butanol involves manipulation of the metabolic networks within bacteria to prioritize the synthesis of the biofuel. Metabolic engineering and genetic engineering tools allow scientists to alter the states of reactions occurring in the organism, utilizing advanced techniques to create a bacterial strain capable of high butanol yield. Optimization can also be accomplished by the transfer of specific genetic information to other uni-cellular species, capitalizing on the traits of multiple organisms to achieve the highest rate of alcohol production.

Using Alternate Carbon Sources

One promising development in biobutanol production technology was discovered in the late summer of 2011—Tulane University's alternative fuel research scientists discovered a strain of *Clostridium*, called "TU-103", that can convert nearly any form of cellulose into butanol, and is the only known strain of *Clostridium*-genus bacteria that can do so in the presence of oxygen. The university's researchers have stated that the source of the "TU-103" *Clostridium* bacteria strain was most likely from the solid waste from one of the plains zebra at New Orleans' Audubon Zoo.

Metabolic engineering can be used to allow an organism to use a cheaper substrate such as glycerol instead of glucose. Because fermentation processes require glucose derived from foods, butanol production can negatively impact food supply. Glycerol is a good alternative source for butanol production. While glucose sources are valuable and limited, glycerol is abundant and has a low market price because it is a waste product of biodiesel production. Butanol production from glycerol is economically viable using metabolic pathways that exist in *Clostridium pasteurianum* bacterium.

A combination of succinate and ethanol can be fermented to produce butyrate (a precursor to butanol fuel) by utilizing the metabolic pathways present in a gram-positive anaerobic bacterium *Clostridium kluyveri*. Succinate is an intermediate of the TCA cycle, which metabolizes glucose. Anaerobic bacteria such as *Clostridium acetobutylicum* and *Clostridium saccharobutylicum* also contain these pathways. Succinate is first activated and then reduced by a two-step reaction to give 4-hydroxybutyrate, which is then metabolized further to crotonyl-coenzyme A (CoA) . Crotonyl-CoA is then converted to butyrate. The genes corresponding to these butanol production pathways from *Clostridium* were cloned to *E. coli.*

In 2012 researchers developed a method for storing electrical energy as chemical energy in higher alcohols (including butanol). These alcohols can then be used as liquid transportation fuels. The team led by James Liao genetically engineered lithoautotrophic microorganism known as *Ralstonia Eutropha* H16 to produce isobutanol and 3-methyl-1-butanol in an electro-bioreactor. Carbon dioxide is the sole carbon source for this process and electricity is used as the energetic component. The process they developed effectively separates the light and dark reactions that occur during photosynthesis. Solar panels are used to convert sunlight to electrical energy which is then converted using the microorganism to a chemical intermediate. The team is now in the process of scaling up the operation and believes this process will be more efficient than the biologic process.

Improving Efficiency

In late 2012, a new discovery made the alternative fuel butanol more attractive to the biofuel industry. Scientist Hao Feng found a method that could significantly reduce the cost of the energy involved in making butanol. His team was able to isolate the butanol molecules during the fermentation process so they do not kill the organisms, and produces 100% or more butanol. After the fermentation process, they used a process called cloud point separation to recover the butanol which used 4 times less energy.

Also in late 2012, utilizing systems metabolic engineering, a Korean research team at the former Korea Advanced Institute of Science and Technology (KAIST) has succeeded in demonstrating an optimized process to increase butanol production by generating an engineered bacterium. Professor Sang Yup Lee at the Department of Chemical and Biomolecular Engineering, KAIST, Dr. Do Young Seung at GS Caltex, a large oil refining company in Korea, and Dr. Yu-Sin Jang at BioFuelChem, a startup butanol company in Korea, applied a systems metabolic engineering approach to improve the production of butanol through enhancing the performance of *Clostridium acetobutylicum*, one of the best known butanol-producing bacteria. In addition, the downstream process was optimized and an in situ recovery process was integrated to achieve higher butanol titer, yield, and productivity. The combination of systems metabolic engineering and bioprocess optimization resulted in the development of a process capable of producing more than 585 g of butanol from 1.8 kg of glucose, which allows the production of this important industrial solvent and advanced biofuel to be cost competitive.

The anaerobic bacteria *C. pasteurianum, C. acetobutylicum*, and other *Clostridium* species have metabolic pathways that convert glycerol to butanol through fermentation. However, the production of butanol from glycerol by fermentation in *C. Pasteurianum* is low. To counter this, a group of researchers used chemical mutagenesis to create a hyper butanol-producing strain. The best mutant strain in this study "MBEL_GLY2" produced 10.8 g of butanol per 80 g of glycerol fed to the bacteria. This improvement compares to the 7.6 g butanol produced by the native bacteria.

Many organisms have the capacity to produce butanol utilizing an acetyl-CoA dependent pathway. The main problem with this pathway is the first reaction involving the condensation of two acetyl-CoA molecules to acetoacetyl-CoA. This reaction is thermodynamically unfavorable due to the positive Gibbs free energy associated with it (dG = 6.8 kcal/mol). Some experimentation has been done that involves increasing the carbon storage through the organism by utilizing carbon dioxide flow through photosynthetic organisms. To follow in this path of research, scientists have attempted to engineer reaction pathways that can enable photosynthetic organisms (like blue-green algae) to produce butanol more efficiently.

A study done by Ethan I. Lan and James C. Liao attempted to utilize the ATP produced during photosynthesis in blue-green algae to work around the thermodynamically unfavorable acetyl-CoA condensation to acetoacetyl-CoA. The native system was re-engineered to have acetyl-CoA react with ATP and CO2 to form an intermediate, malonyl-CoA. Malonyl-CoA then reacts with another acetyl-CoA to form the desired acetoacetyl-CoA. The energy release from ATP hydrolysis (dG = -7.3 kcal/mol) makes this pathway significantly more favorable than standard condensation. Because blue-green algae generate NADPH during photosynthesis, it can be assumed that the cofactor environment is NADPH rich. Therefore, the native reaction pathway was further engineered to use NADPH rather than the standard NADH. All of these adjustments led to a 4-fold increase in butanol production, showing the importance of ATP and cofactor driving forces as a design principle in pathway engineering.

Producers

DuPont and BP plan to make biobutanol the first product of their joint effort to develop, produce, and market next-generation biofuels. In Europe the Swiss company Butalco is developing genetically modified yeasts for the production of biobutanol from cellulosic materials. Gourmet Butanol, a United States-based company, is developing a process that utilizes fungi to convert organic waste into biobutanol.

The number of biobutanol producers with commercial plants coming on line continues to grow monthly. At present, there are number of bioethanol plants, which are being converted to biobutanol plants, which should increase the number of butanol producers as they come on-line.

Distribution

Butanol better tolerates water contamination and is less corrosive than ethanol and more suitable for distribution through existing pipelines for gasoline. In blends with diesel or gasoline, butanol is less likely to separate from this fuel than ethanol if the fuel is contaminated with water. There is also a vapor pressure co-blend synergy with butanol and gasoline containing ethanol, which facilitates ethanol blending. This facilitates storage and distribution of blended fuels.

Properties of Common Fuels

Fuel	Energy density	Air-fuel ratio	Specific energy	Heat of vaporization	RON	MON	AKI
Gasoline and biogasoline	32 MJ/L	14.7	2.9 MJ/kg air	0.36 MJ/kg	91–99	81–89	87-95
Butanol fuel	29.2 MJ/L	11.1	3.2 MJ/kg air	0.43 MJ/kg	96	78	87
Anhydrous Ethanol fuel	19.6 MJ/L	9.0	3.0 MJ/kg air	0.92 MJ/kg	107	89	
Methanol fuel	16 MJ/L	6.4	3.1 MJ/kg air	1.2 MJ/kg	106	92	

Energy Content and Effects on Fuel Economy

Switching a gasoline engine over to butanol would in theory result in a fuel consumption penalty of about 10% but butanol's effect on mileage is yet to be determined by a scientific study. While the energy density for any mixture of gasoline and butanol can be calculated, tests with other alcohol fuels have demonstrated that the effect on fuel economy is not proportional to the change in energy density.

Octane Rating

The octane rating of n-butanol is similar to that of gasoline but lower than that of ethanol and methanol. n-Butanol has a RON (Research Octane number) of 96 and a MON (Motor octane number) of 78 (with a resulting "(R+M)/2 pump octane number" of 87, as used in North America) while t-butanol has octane ratings of 105 RON and 89 MON. t-Butanol is used as an additive in gasoline but cannot be used as a fuel in its pure form because its relatively high melting point of 25.5 °C causes it to gel and solidify near room temperature.

A fuel with a higher octane rating is less prone to knocking (extremely rapid and spontaneous combustion by compression) and the control system of any modern car engine can take advantage of this by adjusting the ignition timing. This will improve energy efficiency, leading to a better fuel economy than the comparisons of energy content different fuels indicate. By increasing the compression ratio, further gains in fuel economy, power and torque can be achieved. Conversely, a fuel with lower octane rating is more prone to knocking and will lower efficiency. Knocking can also cause engine damage. Engines designed to run on 87 octane will not have any additional power/fuel economy from being operated with higher octane fuel.

Air-fuel Ratio

Alcohol fuels, including butanol and ethanol, are partially oxidized and therefore need to run at richer mixtures than gasoline. Standard gasoline engines in cars can adjust the air-fuel ratio to accommodate variations in the fuel, but only within certain limits depending on model. If the limit is exceeded by running the engine on pure ethanol or a gasoline blend with a high percentage of ethanol, the engine will run lean, something which can critically damage components. Compared to ethanol, butanol can be mixed in higher ratios with gasoline for use in existing cars without the need for retrofit as the air-fuel ratio and energy content are closer to that of gasoline.

Specific Energy

Alcohol fuels have less energy per unit weight and unit volume than gasoline. To make it possible to compare the net energy released per cycle a measure called the fuels specific energy is sometimes used. It is defined as the energy released per air fuel ratio. The net energy released per cycle is higher for butanol than ethanol or methanol and about 10% higher than for gasoline.

Viscosity

Substance	Kinematic viscosity at 20 °C
Butanol	3.64 cSt
Diesel	>3 cSt
Ethanol	1.52 cSt
Water	1.0 cSt
Methanol	0.64 cSt
Gasoline	0.4–0.8 cSt

The viscosity of alcohols increase with longer carbon chains. For this reason, butanol is used as an alternative to shorter alcohols when a more viscous solvent is desired. The kinematic viscosity of butanol is several times higher than that of gasoline and about as viscous as high quality diesel fuel.

Heat of Vaporization

The fuel in an engine has to be vaporized before it will burn. Insufficient vaporization is a known problem with alcohol fuels during cold starts in cold weather. As the heat of vaporization of butanol is less than half of that of ethanol, an engine running on butanol should be easier to start in cold weather than one running on ethanol or methanol.

Potential Problems with the Use of Butanol Fuel

The potential problems with the use of butanol are similar to those of ethanol:

- To match the combustion characteristics of gasoline, the utilization of butanol fuel as a substitute for gasoline requires fuel-flow increases (though butanol has only slightly less energy than gasoline, so the fuel-flow increase required is only minimal, maybe 10%, compared to 40% for ethanol.)
- Alcohol-based fuels are not compatible with some fuel system components.
- Alcohol fuels may cause erroneous gas gauge readings in vehicles with capacitance fuel level gauging.
- While ethanol and methanol have lower energy densities than butanol, their higher octane number allows for greater compression ratio and efficiency.
- Butanol is one of many side products produced from current fermentation technologies; as a consequence, current fermentation technologies allow for very low yields of pure extracted butanol. When compared to ethanol, butanol is more fuel efficient as a fuel alternative, but ethanol can be produced at a much lower cost and with much greater yields.
- Butanol is toxic at a rate of 20g per liter and may need to undergo Tier 1 and Tier 2 health effects testing before being permitted as a primary fuel by the EPA.

Possible Butanol Fuel Mixtures

Standards for the blending of ethanol and methanol in gasoline exist in many countries, including the EU, the US and Brazil. Approximate equivalent butanol blends can be calculated from the relations between the stoichiometric fuel-air ratio of butanol, ethanol and gasoline. Common ethanol fuel mixtures for fuel sold as gasoline currently range from 5% to 10%. The share of butanol can be 60% greater than the equivalent ethanol share, which gives a range from 8% to 16%. "Equivalent" in this case refers only to the vehicle's ability to adjust to the fuel. Other properties such as energy density, viscosity and heat of vaporization will vary and may further limit the percentage of butanol that can be blended with gasoline.

Consumer acceptance may be limited due to the potentially offensive banana-like smell of n-butanol. Plans are underway to market a fuel that is 85% Ethanol and 15% Butanol (E85B), so existing E85 internal combustion engines can run on a 100% renewable fuel that could be made without using any fossil fuels. Because its longer hydrocarbon chain causes it to be fairly non-polar, it is more similar to gasoline than it is to ethanol. Butanol has been demonstrated to work in vehicles designed for use with gasoline without modification.

Current Use of Butanol in Vehicles

Currently no production vehicle is known to be approved by the manufacturer for use with 100% butanol. As of early 2009, only few vehicles are approved for even using E85 fuel (i.e. 85% ethanol + 15% gasoline)in the USA. However, in Brazil all vehicle manufacturers (Fiat, Ford, VW, GM, Toyota, Honda, Peugeot, Citroen and others) produce "flex-fuel" vehicles that can run on 100% ethanol or any mix of ethanol and gasoline. These flex fuel cars represent 90% of the sales of personal vehicles in Brazil, in 2009. BP and Dupont, engaged in a joint venture to produce and promote butanol fuel, claim that "biobutanol can be blended up to 10%v/v in European gasoline and 11.5%v/v in US gasoline".

David Ramey drove from Blacklick, Ohio to San Diego, California using 100% butanol in an unmodified 1992 Buick Park Avenue.

In the 2009 Petit Le Mans race, the No. 16 Lola B09/86 - Mazda MZR-R of Dyson Racing ran on a mixture of biobutanol and ethanol developed by team technology partner BP.

Babassu Oil

Babassu palms in Maranhão, Brazil

Babassu oil or cusi oil is a clear light yellow vegetable oil extracted from the seeds of the babassu palm (*Attalea speciosa*), which grows in the Amazon region of South America. It is a non-drying oil used in food, cleaners and skin products. This oil has properties similar to coconut oil and is used in much the same context. It is increasingly being used as a substitute for coconut oil. Babassu oil is about 70% lipids, in the following proportions:

Lauric and myristic acids have melting points relatively close to human body temperature, so babassu oil can be applied to the skin as a solid that melts on contact. This heat transfer can produce a cooling sensation. It is an effective emollient.

During February 2008, a mixture of babassu oil and coconut oil was used to partially power one engine of a Boeing 747, in a biofuel trial sponsored by Virgin Atlantic Airways.

Fatty acid	Percentage
Lauric	50.0%
Myristic	20.0%
Palmitic	11.0%
Oleic	10.0%
Stearic	3.5%

Biogasoline

Biogasoline is gasoline produced from biomass such as algae. Like traditionally produced gasoline, it contains between 6 (hexane) and 12 (dodecane) carbon atoms per molecule and can be used in internal-combustion engines. Biogasoline is chemically different from biobutanol and bioethanol, as these are alcohols, not hydrocarbons.

Companies such as Diversified Energy Corporation are developing approaches to take triglyceride inputs and through a process of deoxygenation and reforming (cracking, isomerizing, aromatizing, and producing cyclic molecules) producing biogasoline. This biogasoline is intended to match the chemical, kinetic, and combustion characteristics of its petroleum counterpart, but with much higher octane levels. Others are pursuing similar approaches based on hydrotreating. And lastly still others are focused on the use of woody biomass for conversion to biogasoline using enzymatic processes.

Structure and Properties

BG100, or 100% biogasoline, can immediately be used as a drop-in substitute for petroleum gasoline in any conventional gasoline engine, and can be distributed in the same fueling infrastructure, as the properties match traditional gasoline from petroleum. Dodecane requires a small percentage of octane booster to match gasoline. Ethanol fuel (E85) requires a special engine and has lower combustion energy and corresponding fuel economy.

But due to biogasoline's chemical similarities it can also be mixed with regular gasoline. You can have higher ratios of biogasoline to gasoline and not have to modify the vehicles engine unlike ethanol.

Comparison to Common Fuels

Fuel	Energy Density MJ/L	Air-fuel ratio	Specific Energy MJ/kg	Heat of Vaporization MJ/kg	RON	MON
Gasoline	34.6	14.6	46.9	0.36	91–99	81–89

Fuel	Energy Density MJ/L	Air-fu-el ratio	Specific Energy MJ/kg	Heat of Vapor-ization MJ/kg	RON	MON
Butanol fuel	29.2	11.2	36.6	0.43	96	78
Ethanol fuel	24.0	9.0	30.0	0.92	129	102
Methanol fuel	19.7	6.5	15.6	1.2	136	104

Production

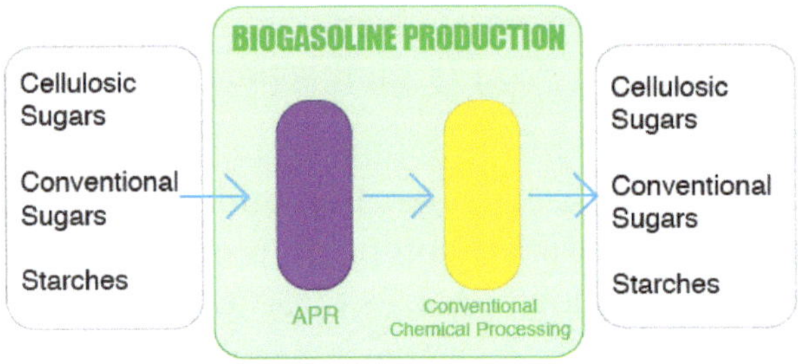

Biogasoline Production Process

Biogasoline is created by turning sugar directly into gasoline. In late March 2010, the world's first biogasoline demonstration plant was started in Madison, WI by Virent Energy Systems, Inc. Virent discovered and developed a technique called Aqueous Phase Reforming (APR) in 2001. APR includes many processes including reforming to generate hydrogen, dehydrogenation of alcohols/hydrogenation of carbonyls, deoxygenation reactions, hydrogenolysis and cyclization. The input for APR is a carbohydrate solution created from plant material, and the product is a mixture of chemicals and oxygenated hydrocarbons. From there, the materials go through further conventional chemical processing to yield the final result: a mixture of non-oxygenated hydrocarbons that the claimed was cost-effective. These hydrocarbons are the exact hydrocarbons found in petroleum fuels which is why today's cars do not need to be altered to run on biogasoline. The only difference is in origin. Petroleum based fuels are made from oil, and biogasoline is made from plants such as beets and sugarcane or cellulosic biomass which would normally be plant waste.

Diesel fuel is made up of linear hydrocarbons. These are long straight carbon atom chains. They differ from the shorter, branched hydrocarbons that make up gasoline. In 2014 Researchers used a feedstock of levulinic acid to create biogasoline. Levulinic acid is derived from cellulose material, such as corn stalks, straw or other plant waste. That waste does not have to be fermented. The fuel-making process is reportedly inexpensive and offers yields of over 60 percent.

Research

Research is conducted in both the academic and private sectors.

Academic

Virginia Polytechnic Institute and State University has been researching for the past four years on making stable biogasoline in current oil refineries. Their focus of the research was the length of time bio-oil's shelf-life. The use of catalysts was used in order to remove impurities from the processed plant sugars. The researchers extended the time from three months to over a year.

Iowa State University researchers use a type of fermentation in their research. They first start by forming a gaseous mixture and pyrolysize it. The result of the pyrolysis is bio-oil which the sugar rich portion is fermented and distilled to create water and ethanol. But the high acetate portion is then separated into biogasoline, water, and biomass.

Private

Virent Energy Systems, Inc. which is located in Madison, Wisconsin in conjunction with Shell has developed a technique to turn plant sugars from wheat straw, corn stalks, and sugarcane pulp into biogasoline. The sugars are converted into hydrocarbons similar to those in regular gasoline by the use of catalysts.

Economic Viability and Future

One of the major problems facing the economic viability of biogasoline is the high up-front cost. Research groups are finding that current investment groups are impatient with the pace of biogasoline progress. In addition, environmental groups may demand that biogasoline that is produced in a way that protects wildlife, especially fish. A research group studying the economic viability of biofuels found that current techniques of production and high costs of production will prevent biogasoline from being accessible to the general public. The group determined that the price of biogasoline would need to be approximately $800 per barrel, which they determine as unlikely with current production costs. Another problem inhibiting the success of biogasoline is the lack of tax relief. The government is providing tax relief for ethanol fuels but has yet to offer tax relief for biogasoline. This makes biogasoline a much less attractive option to consumers. Lastly, producing biogasoline could have a large effect on the farming industry. If biogasoline became a serious alternative, a large percentage of our existing arable land would be converted to grow crops solely for biogasoline. This could decrease the amount of land used to farm food for human consumption and may decrease overall feedstock. This would cause an increase in overall food cost.

While there may be some problems facing the economic viability of biogasoline, the partnership between Royal Dutch Shell and Virent Energy Systems, Inc., a bioscience firm based in Madison, WI, to further research biogasoline is an encouraging sign for biogasoline's future. In addition, many nations are enacting policies that increase the use of biogasoline within the country to help curb the cost of fossil fuels and create more energy independence. Current efforts by the partnership are focused on improving the technology and making it available for large-scale production.

Methanol Fuel

Methanol is an alternative fuel for internal combustion and other engines, either in combination with gasoline or directly ("neat"). It is used in racing cars in many countries. In the U.S., methanol fuel has received less attention than ethanol fuel as an alternative to petroleum-based fuels, because in the 2000s particularly, the support of corn-based ethanol offered certain political advantages. In general, ethanol is less toxic and has higher energy density, although methanol is less expensive to produce sustainably and is a less expensive way to reduce the carbon footprint. However, for optimizing engine performance, fuel availability, toxicity and political advantage, a blend of ethanol, methanol and petroleum is likely to be preferable to using any of these individual substances alone. Methanol may be made from hydrocarbon or renewable resources, in particular natural gas and biomass respectively. It can also be synthesized from CO_2 (carbon dioxide) and hydrogen.

History and Production

Historically, methanol was first produced by destructive distillation (pyrolysis) of wood, resulting in its common English name of wood alcohol.

At present, methanol is usually produced using methane (the chief constituent of natural gas) as a raw material. In China, methanol is made for fuel from coal.

"Biomethanol" may be produced by gasification of organic materials to synthesis gas followed by conventional methanol synthesis. This route can offer methanol production from biomass at efficiencies up to 75%. Widespread production by this route has a proposed potential to offer methanol fuel at a low cost and with benefits to the environment. These production methods, however, are not suitable for small-scale production.

Recently, methanol fuel has been produced using renewable energy and carbon dioxide as a feedstock. Carbon Recycling International, an Icelandic-American company, completed the first commercial scale renewable methanol plant in 2011.

Major Fuel Use

During the OPEC 1973 oil crisis, Reed and Lerner (1973) proposed methanol from coal as a proven fuel with well-established manufacturing technology and sufficient resources to replace gasoline. Hagen (1976) reviewed prospects for synthesizing methanol from renewable resources and its use as a fuel. Then in 1986, the Swedish Motor Fuel Technology Co. (SBAD) extensively reviewed the use of alcohols and alcohol blends as motor fuels. It reviewed the potential for methanol production from natural gas, very heavy oils, bituminous shales, coals, peat and biomass. In 2005, 2006 Nobel prize winner George A. Olah and colleagues advocated an entire methanol economy based on energy storage in synthetically produced methanol. The Methanol Institute, the methanol trade industry organization, posts reports and presentations on methanol. Director Gregory Dolan presented the 2008 global methanol fuel industry in China.

On January 26, 2011, the European Union's Directorate-General for Competition approved the Swedish Energy Agency's award of 500 million Swedish kronor (approx. €56M as at January 2011) toward the construction of a 3 billion Swedish kronor (approx. €335M) industrial scale experimental development biofuels plant for production of Biomethanol and BioDME at the Domsjö Fabriker biorefinery complex in Örnsköldsvik, Sweden, using Chemrec's black liquor gasification technology.

Uses

Internal Combustion Engine Fuel

Both methanol and ethanol burn at lower temperatures than gasoline, and both are less volatile, making engine starting in cold weather more difficult. Using methanol as a fuel in spark-ignition engines can offer an increased thermal efficiency and increased power output (as compared to gasoline) due to its high octane rating (114) and high heat of vaporization. However, its low energy content of 19.7 MJ/kg and stoichiometric air-to-fuel ratio of 6.42:1 mean that fuel consumption (on volume or mass bases) will be higher than hydrocarbon fuels. The extra water produced also makes the charge rather wet (similar to hydrogen/oxygen combustion engines) and with the formation of acidic products during combustion, the wearing of valves, valve seats and cylinder might be higher than with hydrocarbon burning. Certain additives may be added to the fuel in order to neutralize these acids.

Methanol, just like ethanol, contains soluble and insoluble contaminants. These soluble contaminants, halide ions such as chloride ions, have a large effect on the corrosivity of alcohol fuels. Halide ions increase corrosion in two ways; they chemically attack passivating oxide films on several metals causing pitting corrosion, and they increase the conductivity of the fuel. Increased electrical conductivity promotes electric, galvanic, and ordinary corrosion in the fuel system. Soluble contaminants, such as aluminum hydroxide, itself a product of corrosion by halide ions, clog the fuel system over time.

Methanol is (In automotive terms) hygroscopic, meaning it will absorb water vapor directly from the atmosphere. Because absorbed water dilutes the fuel value of the methanol (although it suppresses engine knock), and may cause phase separation of methanol-gasoline blends, containers of methanol fuels must be kept tightly sealed.

Racing

Beginning in 1965, pure methanol was used widespread in USAC Indy car competition, which at the time included the Indianapolis 500.

A seven-car crash on the second lap of the 1964 Indianapolis 500 resulted in USAC's decision to encourage, and later mandate, the use of methanol. Eddie Sachs and Dave MacDonald died in the crash when their gasoline-fueled cars exploded. The gasoline-triggered fire created a dangerous cloud of thick black smoke that completely blocked the view of the track for oncoming cars. Johnny Rutherford, one of the other drivers involved, drove a methanol-fueled car, which also leaked following the crash. While this car burned from the impact of the first fireball, it formed a much smaller inferno than the gasoline cars, and one that burned invisibly. That testimony, and pressure from Indianapolis Star writer George Moore, led to the switch to alcohol fuel in 1965.

Methanol was used by the CART circuit during its entire campaign (1979–2007). It is also used by many-short track organizations, especially midget, sprint cars and speedway bikes. Pure methanol was used by the IRL from 1996-2006.

In 2006, in partnership with the ethanol industry, the IRL used a mixture of 10% ethanol and 90% methanol as its fuel. Starting in 2007, the IRL switched to "pure" ethanol, E100.

Methanol fuel is also used extensively in drag racing, primarily in the Top Alcohol category, while between 10% and 20% methanol may be used in Top Fuel classes in addition to Nitromethane.

Formula One racing continues to use gasoline as its fuel, but in prewar grand prix racing methanol was often used in the fuel.

Methanol is also used in Monster Truck racing.

Fuel for Model Engines

The earliest model engines for free-flight model aircraft flown before the end of World War II used a 3:1 mix of white gas and heavy viscosity motor oil for the two-stroke spark ignition engines used for the hobby at that time. By 1948, the new glow plug-ignition model engines began to take over the market, requiring the use of methanol fuel to react in a catalytic reaction with the coiled platinum filament in a glow plug for the engine to run, usually using a castor oil-based lubricant contained in the fuel mix at about a 4:1 ratio. The glow-ignition variety of model engine, because it no longer required an onboard battery, ignition coil, ignition points and capacitor that a spark ignition model engine required, saved valuable weight and allowed model aircraft to have better flight performance. In their traditionally popular two-stroke and increasingly popular four-stroke forms, currently produced single cylinder methanol-fueled glow engines are the usual choice for radio controlled aircraft for recreational use, for engine sizes that can range from 0.8 cm3 (0.049 cu.in.) to as large as 25 to 32 cm3 (1.5-2.0 cu.in) displacement, and significantly larger displacements for twin and multi-cylinder opposed-cylinder and radial configuration model aircraft engines, many of which are of four-stroke configuration. Most methanol-fueled model engines, especially those made outside North America, can easily be run on so-called *FAI*-specification methanol fuel. Such fuel mixtures can be required by the FAI for certain events in so-called FAI "Class F" international competition, that forbid the use of nitromethane as a glow engine fuel component. In contrast, firms in North America that make methanol-fueled model engines, or who are based outside that continent and have a major market in North America for such miniature powerplants, tend to produce engines that can and often do run best with a certain percentage of nitromethane in the fuel, which when used can be as little as 5% to 10% of volume, and can be as much as 25 to 30% of the total fuel volume.

Toxicity

Methanol occurs naturally in the human body and in some fruits, but is poisonous in high concentration. Ingestion of 10 ml can cause blindness and 60-100 ml can be fatal if the condition is untreated. Like many volatile chemicals, methanol does not have to be swallowed to be dangerous since the liquid can be absorbed through the skin, and the vapors through the lungs. Methanol fuel is much safer when blended with ethanol, even at relatively low ethanol percentages.

US maximum allowed exposure in air (40 h/week) is 1900 mg/m^3 for ethanol, 900 mg/m^3 for gasoline, and 1260 mg/m^3 for methanol. However, it is much less volatile than gasoline, and therefore has lower evaporative emissions, producing a lower exposure risk for an equivalent spill. While methanol offers somewhat different toxicity exposure pathways, the effective toxicity is no worse than those of benzene or gasoline, and methanol poisoning is far easier to treat successfully. One substantial concern is that methanol poisoning generally must be treated while it is still asymptomatic for full recovery.

Inhalation risk is mitigated by a characteristic pungent odor. At concentrations greater than 2,000 ppm (0.2%) it is generally quite noticeable, however lower concentrations may remain undetected while still being potentially toxic over longer exposures, and may still present a fire/explosion hazard. Again, this is similar to gasoline and ethanol; standard safety protocols exist for methanol and are very similar to those for gasoline and ethanol.

Use of methanol fuel reduces the exhaust emissions of certain hydrocarbon-related toxins such as benzene and 1,3 butadiene, and dramatically reduces long term groundwater pollution caused by fuel spills. Unlike benzene-family fuels, methanol will rapidly and non-toxically biodegrade with no long-term harm to the environment as long as it is sufficiently diluted.

Fire Safety

Methanol is far more difficult to ignite than gasoline and burns about 60% slower. A methanol fire releases energy at around 20% of the rate of a gasoline fire, resulting in a much cooler flame. This results in a much less dangerous fire that is easier to contain with proper protocols. Unlike gasoline fires, water is acceptable and even preferred as a fire suppressant for methanol fires, since this both cools the fire and rapidly dilutes the fuel below the concentration where it will maintain self-flammability. These facts mean that, as a vehicle fuel, methanol has great safety advantages over gasoline. Ethanol shares many of these same advantages.

Since methanol vapor is heavier than air, it will linger close to the ground or in a pit unless there is good ventilation, and if the concentration of methanol is above 6.7% in air it can be lit by a spark and will explode above 54 F / 12 C. Once ablaze, an undiluted methanol fire gives off very little visible light, making it potentially very hard to see the fire or even estimate its size in bright daylight, although in the vast majority of cases, existing pollutants or flammables in the fire (such as tires or asphalt) will color and enhance the visibility of the fire. Ethanol, natural gas, hydrogen, and other existing fuels offer similar fire-safety challenges, and standard safety and firefighting protocols exist for all such fuels.

Post-accident environmental damage mitigation is facilitated by the fact that low-concentration methanol is biodegradable, of low toxicity, and non-persistent in the environment. Post-fire cleanup often merely requires large additional amounts of water to dilute the spilled methanol followed by vacuuming or absorption recovery of the fluid. Any methanol that unavoidably escapes into the environment will have little long-term impact, and with sufficient dilution will rapidly biodegrade with little to no environmental damage due to toxicity. A methanol spill that combines with an existing gasoline spill can cause the mixed methanol/gasoline spill to persist about 30% to 35% longer than the gasoline alone would have done.

Use

United States

The State of California ran an experimental program from 1980 to 1990 that allowed anyone to convert a gasoline vehicle to 85% methanol with 15% additives of choice. Over 500 vehicles were converted to high compression and dedicated use of the 85/15 methanol and ethanol, with great results. Detroit was not willing to produce any methanol or ethanol vehicles without government subsidy.

In 1982 the big three were each given $5,000,000 for design and contracts for 5,000 vehicles to be bought by the State. That was the beginning of the low-compression flexible-fuel vehicles that we can still buy today.

In 2005, California's Governor, Arnold Schwarzenegger, stopped the use of methanol to join the expanding use of ethanol driven by producers of corn. Ethanol is currently (as of 2007) priced at 3 to 4 dollars per gallon (0.8 to 1.05 dollars per liter) at the pump, while methanol made from natural gas remains at 47 cents per gallon (12.5 cents per liter) in bulk, not at the pump.

Presently there are no operating gas stations in California supplying methanol in their pumps. Rep. Eliot Engel [D-NY17] has introduced "An Open Fuel Standard" Act in Congress: "To require automobile manufacturers to ensure that not less than 80 percent of the automobiles manufactured or sold in the United States by each such manufacturer to operate on fuel mixtures containing 85 percent ethanol, 85 percent methanol, or biodiesel."

European Union

The amended Fuel Quality Directive adopted in 2009 allows up to 3% v/v blend-in of methanol in petrol.

Brazil

A drive to add an appreciable percentage of methanol to gasoline got very close to implementation in Brazil, following a pilot test set up by a group of scientists involving blending gasoline with methanol between 1989 and 1992. The larger-scale pilot experiment that was to be conducted in São Paulo was vetoed at the last minute by the city's mayor, out of concern for the health of gas station workers, who are mostly illiterate and could not be expected to follow safety precautions. As of 2006, the idea has not resurfaced.

Biogas

Pipes carrying biogas (foreground), natural gas and condensate

Biogas typically refers to a mixture of different gases produced by the breakdown of organic matter in the absence of oxygen. Biogas can be produced from raw materials such as agricultural waste, manure, municipal waste, plant material, sewage, green waste or food waste. Biogas is a renewable energy source and in many cases exerts a very small carbon footprint.

Biogas can be produced by anaerobic digestion with anaerobic organisms, which digest material inside a closed system, or fermentation of biodegradable materials.

Biogas is primarily methane (CH4) and carbon dioxide (CO_2) and may have small amounts of hydrogen sulfide (H2S), moisture and siloxanes. The gases methane, hydrogen, and carbon monoxide (CO) can be combusted or oxidized with oxygen. This energy release allows biogas to be used as a fuel; it can be used for any heating purpose, such as cooking. It can also be used in a gas engine to convert the energy in the gas into electricity and heat.

Biogas can be compressed, the same way natural gas is compressed to CNG, and used to power motor vehicles. In the UK, for example, biogas is estimated to have the potential to replace around 17% of vehicle fuel. It qualifies for renewable energy subsidies in some parts of the world. Biogas can be cleaned and upgraded to natural gas standards, when it becomes bio-methane. Biogas is considered to be a renewable resource because its production-and-use cycle is continuous, and it generates no net carbon dioxide. Organic material grows, is converted and used and then regrows in a continually repeating cycle. From a carbon perspective, as much carbon dioxide is absorbed from the atmosphere in the growth of the primary bio-resource as is released when the material is ultimately converted to energy.

Production

Biogas production in rural Germany

Biogas is produced as landfill gas (LFG), which is produced by the breakdown of Biodegradable waste inside a landfill due to chemical reactions and microbes, or as digested gas, produced inside an anaerobic digester. A *biogas plant* is the name often given to an anaerobic digester that treats farm wastes or energy crops. It can be produced using anaerobic digesters (air-tight tanks with different configurations). These plants can be fed with energy crops such as maize silage or biodegradable wastes including sewage sludge and food waste. During the process, the microorganisms transform biomass waste into biogas (mainly methane and carbon dioxide) and digestate. The biogas is a renewable energy that can be used for heating, electricity, and many other operations that use a reciprocating internal combustion engine, such as GE Jenbacher or Caterpillar gas engines. Other internal combustion engines such as gas turbines are suitable for the conversion of biogas into both electricity and heat. The digestate is the remaining inorganic matter that was not transformed into biogas. It can be used as an agricultural fertiliser.

There are two key processes: mesophilic and thermophilic digestion which is dependent on temperature. In experimental work at University of Alaska Fairbanks, a 1000-litre digester using psychrophiles harvested from "mud from a frozen lake in Alaska" has produced 200–300 liters of methane per day, about 20%–30% of the output from digesters in warmer climates.

Dangers

The dangers of biogas are mostly similar to those of natural gas, but with an additional risk from the toxicity of its hydrogen sulfide fraction. Biogas can be explosive when mixed in the ratio of one part biogas to 8-20 parts air. Special safety precautions have to be taken for entering an empty biogas digester for maintenance work.

It is important that a biogas system never has negative pressure as this could cause an explosion. Negative gas pressure can occur if too much gas is removed or leaked; Because of this biogas should not be used at pressures below one column inch of water, measured by a pressure gauge.

Frequent smell checks must be performed on a biogas system. If biogas is smelled anywhere windows and doors should be opened immediately. If there is a fire the gas should be shut off at the gate valve of the biogas system.

Landfill Gas

Landfill gas is produced by wet organic waste decomposing under anaerobic conditions in a biogas.

The waste is covered and mechanically compressed by the weight of the material that is deposited above. This material prevents oxygen exposure thus allowing anaerobic microbes to thrive. This gas builds up and is slowly released into the atmosphere if the site has not been engineered to capture the gas. Landfill gas released in an uncontrolled way can be hazardous since it can become explosive when it escapes from the landfill and mixes with oxygen. The lower explosive limit is 5% methane and the upper is 15% methane.

The methane in biogas is 20 times more potent a greenhouse gas than carbon dioxide. Therefore, uncontained landfill gas, which escapes into the atmosphere may significantly contribute to the effects of global warming. In addition, volatile organic compounds (VOCs) in landfill gas contribute to the formation of photochemical smog.

Technical

Biochemical oxygen demand (BOD) is a measure of the amount of oxygen required by aerobic micro-organisms to decompose the organic matter in a sample of water. Knowing the energy density of the material being used in the biodigester as well as the BOD for the liquid discharge allows for the calculation of the daily energy output from a biodigester.

Another term related to biodigesters is effluent dirtiness, which tells how much organic material there is per unit of biogas source. Typical units for this measure are in mg BOD/litre. As an example, effluent dirtiness can range between 800–1200 mg BOD/litre in Panama.

From 1 kg of decommissioned kitchen bio-waste, 0.45 m^3 of biogas can be obtained. The price for collecting biological waste from households is approximately €70 per ton.

Composition

Typical composition of biogas		
Compound	**Formula**	**%**
Methane	CH 4	50–75
Carbon dioxide	CO 2	25–50
Nitrogen	N 2	0–10
Hydrogen	H 2	0–1
Hydrogen sulfide	H 2S	0–3
Oxygen	O 2	0–0.5
Source: *www.kolumbus.fi, 2007*		

The composition of biogas varies depending upon the origin of the anaerobic digestion process. Landfill gas typically has methane concentrations around 50%. Advanced waste treatment technologies can produce biogas with 55%–75% methane, which for reactors with free liquids can be increased to 80%-90% methane using in-situ gas purification techniques. As produced, biogas contains water vapor. The fractional volume of water vapor is a function of biogas temperature; correction of measured gas volume for water vapor content and thermal expansion is easily done via simple mathematics which yields the standardized volume of dry biogas.

In some cases, biogas contains siloxanes. They are formed from the anaerobic decomposition of materials commonly found in soaps and detergents. During combustion of biogas containing siloxanes, silicon is released and can combine with free oxygen or other elements in the combustion gas. Deposits are formed containing mostly silica (SiO_2) or silicates (Si_xO_y) and can contain calcium, sulfur, zinc, phosphorus. Such *white mineral* deposits accumulate to a surface thickness of several millimeters and must be removed by chemical or mechanical means.

Practical and cost-effective technologies to remove siloxanes and other biogas contaminants are available.

For 1000 kg (wet weight) of input to a typical biodigester, total solids may be 30% of the wet weight while volatile suspended solids may be 90% of the total solids. Protein would be 20% of the volatile solids, carbohydrates would be 70% of the volatile solids, and finally fats would be 10% of the volatile solids.

Benefits of Manure Derived Biogas

High levels of methane are produced when manure is stored under anaerobic conditions. During storage and when manure has been applied to the land, nitrous oxide is also produced as a byproduct of the denitrification process. Nitrous oxide (N_2O) is 320 times more aggressive as a greenhouse gas than carbon dioxide and methane 25 times more than carbon dioxide.

By converting cow manure into methane biogas via anaerobic digestion, the millions of cattle in the United States would be able to produce 100 billion kilowatt hours of electricity, enough to power millions of homes across the United States. In fact, one cow can produce enough manure in one day to generate 3 kilowatt hours of electricity; only 2.4 kilowatt hours of electricity are needed to power a single 100-watt light bulb for one day. Furthermore, by converting cattle manure into methane biogas instead of letting it decompose, global warming gases could be reduced by 99 million metric tons or 4%.

Applications

Biogas can be used for electricity production on sewage works, in a CHP gas engine, where the waste heat from the engine is conveniently used for heating the digester; cooking; space heating; water heating; and process heating. If compressed, it can replace compressed natural gas for use in vehicles, where it can fuel an internal combustion engine or fuel cells and is a much more effective displacer of carbon dioxide than the normal use in on-site CHP plants.

A biogas bus in Linköping, Sweden

Biogas Upgrading

Raw biogas produced from digestion is roughly 60% methane and 29% CO_2 with trace elements of H_2S; it is not of high enough quality to be used as fuel gas for machinery. The corrosive nature of H_2S alone is enough to destroy the internals of a plant.

Methane in biogas can be concentrated via a biogas upgrader to the same standards as fossil natural gas, which itself has to go through a cleaning process, and becomes *biomethane*. If the local gas network allows, the producer of the biogas may use their distribution networks. Gas must be very clean to reach pipeline quality and must be of the correct composition for the distribution network to accept. Carbon dioxide, water, hydrogen sulfide, and particulates must be removed if present.

There are four main methods of upgrading: water washing, pressure swing adsorption, selexol adsorption, and amine gas treating. In addition to these, the use of membrane separation technology for biogas upgrading is increasing, and there are already several plants operating in Europe and USA.

The most prevalent method is water washing where high pressure gas flows into a column where the carbon dioxide and other trace elements are scrubbed by cascading water running counter-flow to the gas. This arrangement could deliver 98% methane with manufacturers guaranteeing maximum 2% methane loss in the system. It takes roughly between 3% and 6% of the total energy output in gas to run a biogas upgrading system.

Biogas Gas-grid Injection

Gas-grid injection is the injection of biogas into the methane grid (natural gas grid). Injections includes biogas until the breakthrough of micro combined heat and power two-thirds of all the energy produced by biogas power plants was lost (the heat), using the grid to transport the gas to customers, the electricity and the heat can be used for on-site generation resulting in a reduction of losses in the transportation of energy. Typical energy losses in natural gas transmission systems range from 1% to 2%. The current energy losses on a large electrical system range from 5% to 8%.

Biogas in Transport

"Biogaståget Amanda" ("The Biogas Train Amanda") train near Linköping station, Sweden

If concentrated and compressed, it can be used in vehicle transportation. Compressed biogas is becoming widely used in Sweden, Switzerland, and Germany. A biogas-powered train, named Biogaståget Amanda (The Biogas Train Amanda), has been in service in Sweden since 2005. Biogas powers automobiles. In 1974, a British documentary film titled *Sweet as a Nut* detailed the biogas production process from pig manure and showed how it fueled a custom-adapted combustion engine. In 2007, an estimated 12,000 vehicles were being fueled with upgraded biogas worldwide, mostly in Europe.

Measuring in Biogas Environments

Biogas is part of the wet gas and condensing gas (or air) category that includes mist or fog in the gas stream. The mist or fog is predominately water vapor that condenses on the sides of pipes or stacks throughout the gas flow. Biogas environments include wastewater digesters, landfills, and animal feeding operations (covered livestock lagoons).

Ultrasonic flow meters are one of the few devices capable of measuring in a biogas atmosphere. Most thermal flow meters are unable to provide reliable data because the moisture causes steady high flow readings and continuous flow spiking, although there are single-point insertion thermal mass flow meters capable of accurately monitoring biogas flows with minimal pressure drop. They can handle moisture variations that occur in the flow stream because of daily and seasonal temperature fluctuations, and account for the moisture in the flow stream to produce a dry gas value.

Legislation

European Union

The European Union has legislation regarding waste management and landfill sites called the Landfill Directive.

Countries such as the United Kingdom and Germany now have legislation in force that provides farmers with long-term revenue and energy security.

United States

The United States legislates against landfill gas as it contains VOCs. The United States Clean Air Act and Title 40 of the Code of Federal Regulations (CFR) requires landfill owners to estimate the quantity of non-methane organic compounds (NMOCs) emitted. If the estimated NMOC emissions exceeds 50 tonnes per year, the landfill owner is required to collect the gas and treat it to remove the entrained NMOCs. Treatment of the landfill gas is usually by combustion. Because of the remoteness of landfill sites, it is sometimes not economically feasible to produce electricity from the gas.

Global Developments

United States

With the many benefits of biogas, it is starting to become a popular source of energy and is starting to be used in the United States more. In 2003, the United States consumed 147 trillion BTU of energy from "landfill gas", about 0.6% of the total U.S. natural gas consumption. Methane biogas derived from cow manure is being tested in the U.S. According to a 2008 study, collected by the *Science and Children* magazine, methane biogas from cow manure would be sufficient to produce 100 billion kilowatt hours enough to power millions of homes across America. Furthermore, methane biogas has been tested to prove that it can reduce 99 million metric tons of greenhouse gas emissions or about 4% of the greenhouse gases produced by the United States.

In Vermont, for example, biogas generated on dairy farms was included in the CVPS Cow Power program. The program was originally offered by Central Vermont Public Service Corporation as a voluntary tariff and now with a recent merger with Green Mountain Power is now the GMP Cow Power Program. Customers can elect to pay a premium on their electric bill, and that premium is passed directly to the farms in the program. In Sheldon, Vermont, Green Mountain Dairy has provided renewable energy as part of the Cow Power program. It started when the brothers who own the farm, Bill and Brian Rowell, wanted to address some of the manure management challenges faced by dairy farms, including manure odor, and nutrient availability for the crops they need to grow to feed the animals. They installed an anaerobic digester to process the cow and milking center waste from their 950 cows to produce renewable energy, a bedding to replace sawdust, and a plant-friendly fertilizer. The energy and environmental attributes are sold to the GMP Cow Power program. On average, the system run by the Rowells produces enough electricity to power 300 to 350 other homes. The generator capacity is about 300 kilowatts.

In Hereford, Texas, cow manure is being used to power an ethanol power plant. By switching to methane biogas, the ethanol power plant has saved 1000 barrels of oil a day. Over all, the power plant has reduced transportation costs and will be opening many more jobs for future power plants that will rely on biogas.

In Oakley, Kansas, an ethanol plant considered to be one of the largest biogas facilities in North America is using Integrated Manure Utilization System "IMUS" to produce heat for its boilers by utilizing feedlot manure, municipal organics and ethanol plant waste. At full capacity the plant is expected to replace 90% of the fossil fuel used in the manufacturing process of ethanol.

Europe

The level of development varies greatly in Europe. While countries such as Germany, Austria and Sweden are fairly advanced in their use of biogas, there is a vast potential for this renewable energy source in the rest of the continent, especially in Eastern Europe. Different legal frameworks, education schemes and the availability of technology are among the prime reasons behind this untapped potential. Another challenge for the further progression of biogas has been negative public perception.

In February 2009, the European Biogas Association (EBA) was founded in Brussels as a non-profit organisation to promote the deployment of sustainable biogas production and use in Europe. EBA's strategy defines three priorities: establish biogas as an important part of Europe's energy mix, promote source separation of household waste to increase the gas potential, and support the production of biomethane as vehicle fuel. In July 2013, it had 60 members from 24 countries across Europe.

UK

As of September 2013, there are about 130 non-sewage biogas plants in the UK. Most are on-farm, and some larger facilities exist off-farm, which are taking food and consumer wastes.

On 5 October 2010, biogas was injected into the UK gas grid for the first time. Sewage from over 30,000 Oxfordshire homes is sent to Didcot sewage treatment works, where it is treated in an anaerobic digestor to produce biogas, which is then cleaned to provide gas for approximately 200 homes.

In 2015 the Green-Energy company Ecotricity announced their plans to build three grid-injecting digester's.

Germany

Germany is Europe's biggest biogas producer and the market leader in biogas technology. In 2010 there were 5,905 biogas plants operating throughout the country: Lower Saxony, Bavaria, and the eastern federal states are the main regions. Most of these plants are employed as power plants. Usually the biogas plants are directly connected with a CHP which produces electric power by burning the bio methane. The electrical power is then fed into the public power grid. In 2010, the total installed electrical capacity of these power plants was 2,291 MW. The electricity supply was approximately 12.8 TWh, which is 12.6% of the total generated renewable electricity.

Biogas in Germany is primarily extracted by the co-fermentation of energy crops (called 'NawaRo', an abbreviation of *nachwachsende Rohstoffe*, German for renewable resources) mixed with manure. The main crop used is corn. Organic waste and industrial and agricultural residues such as waste from the food industry are also used for biogas generation.In this respect, biogas production in Germany differs significantly from the UK, where biogas generated from landfill sites is most common.

Biogas production in Germany has developed rapidly over the last 20 years. The main reason is the legally created frameworks. Government support of renewable energy started in 1991 with the Electricity Feed-in Act (*StrEG*). This law guaranteed the producers of energy from renewable sources the feed into the public power grid, thus the power companies were forced to take all produced energy from independent private producers of green energy. In 2000 the Electricity Feed-in Act was replaced by the Renewable Energy Sources Act (*EEG*). This law even guaranteed a fixed compensation for the produced electric power over 20 years. The amount of around 8 ¢/kWh gave farmers the opportunity to become energy suppliers and gain a further source of income.

The German agricultural biogas production was given a further push in 2004 by implementing the so-called NawaRo-Bonus. This is a special payment given for the use of renewable resources, that is, energy crops. In 2007 the German government stressed its intention to invest further effort and support in improving the renewable energy supply to provide an answer on growing climate challenges and increasing oil prices by the 'Integrated Climate and Energy Programme'.

This continual trend of renewable energy promotion induces a number of challenges facing the management and organisation of renewable energy supply that has also several impacts on the biogas production. The first challenge to be noticed is the high area-consuming of the biogas electric power supply. In 2011 energy crops for biogas production consumed an area of circa 800,000 ha in Germany. This high demand of agricultural areas generates new competitions with the food industries that did not exist hitherto. Moreover, new industries and markets were created in predominately rural regions entailing different new players with an economic, political and civil background. Their influence and acting has to be governed to gain all advantages this new source of energy is offering. Finally biogas will furthermore play an important role in the German renewable energy supply if good governance is focused.

Subcontinent

Biogas in India has been traditionally based on dairy manure as feed stock and these "gobar" gas plants have been in operation for a long period of time, especially in rural India. In the last 2-3 decades, research organisations with a focus on rural energy security have enhanced the design of the systems resulting in newer efficient low cost designs such as the Deenabandhu model.

The Deenabandhu Model is a new biogas-production model popular in India. (*Deenabandhu* means "friend of the helpless.") The unit usually has a capacity of 2 to 3 cubic metres. It is constructed using bricks or by a ferrocement mixture. In India, the brick model costs slightly more than the ferrocement model; however, India's Ministry of New and Renewable Energy offers some subsidy per model constructed.

LPG (Liquefied Petroleum Gas) is a key source of cooking fuel in urban India and its prices have been increasing along with the global fuel prices. Also the heavy subsidies provided by the successive governments in promoting LPG as a domestic cooking fuel has become a financial burden renewing the focus on biogas as a cooking fuel alternative in urban establishments. This has led to the development of prefabricated digester for modular deployments as compared to RCC and cement structures which take a longer duration to construct. Renewed focus on process technology like the Biourja process model has enhanced the stature of medium and large scale anaerobic digester in India as a potential alternative to LPG as primary cooking fuel.

In India, Nepal, Pakistan and Bangladesh biogas produced from the anaerobic digestion of manure in small-scale digestion facilities is called gobar gas; it is estimated that such facilities exist in over 2 million households in India, 50,000 in Bangladesh and thousands in Pakistan, particularly North Punjab, due to the thriving population of livestock. The digester is an airtight circular pit made of concrete with a pipe connection. The manure is directed to the pit, usually straight from the cattle shed. The pit is filled with a required quantity of wastewater. The gas pipe is connected to the kitchen fireplace through control valves. The combustion of this biogas has very little odour or smoke. Owing to simplicity in implementation and use of cheap raw materials in villages, it is one of the most environmentally sound energy sources for rural needs. One type of these system is the Sintex Digester. Some designs use vermiculture to further enhance the slurry produced by the biogas plant for use as compost.

To create awareness and associate the people interested in biogas, the Indian Biogas Association was formed. It aspires to be a unique blend of nationwide operators, manufacturers and planners of biogas plants, and representatives from science and research. The association was founded in 2010 and is now ready to start mushrooming. Its motto is "propagating Biogas in a sustainable way".

In Pakistan, the Rural Support Programmes Network is running the Pakistan Domestic Biogas Programme which has installed 5,360 biogas plants and has trained in excess of 200 masons on the technology and aims to develop the Biogas Sector in Pakistan.

In Nepal, the government provides subsidies to build biogas plant.

China

The Chinese had experimented the applications of biogas since 1958. Around 1970, China had installed 6,000,000 digesters in an effort to make agriculture more efficient. During the last years the technology has met high growth rates. This seems to be the earliest developments in generating biogas from agricultural waste.

In Developing Nations

Domestic biogas plants convert livestock manure and night soil into biogas and slurry, the fermented manure. This technology is feasible for small-holders with livestock producing 50 kg manure per day, an equivalent of about 6 pigs or 3 cows. This manure has to be collectable to mix it with water and feed it into the plant. Toilets can be connected. Another precondition is the temperature that affects the fermentation process. With an optimum at 36 C° the technology especially applies for those living in a (sub) tropical climate. This makes the technology for small holders in developing countries often suitable.

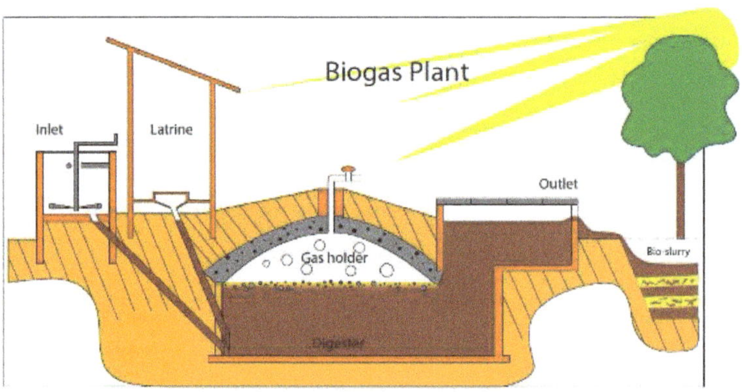

Simple sketch of household biogas plant

Depending on size and location, a typical brick made fixed dome biogas plant can be installed at the yard of a rural household with the investment between US$300 to $500 in Asian countries and up to $1400 in the African context. A high quality biogas plant needs minimum maintenance costs and can produce gas for at least 15–20 years without major problems and re-investments. For the user, biogas provides clean cooking energy, reduces indoor air pollution, and reduces the time needed for traditional biomass collection, especially for women and children. The slurry is a clean organic fertilizer that potentially increases agricultural productivity.

Domestic biogas technology is a proven and established technology in many parts of the world, especially Asia. Several countries in this region have embarked on large-scale programmes on domestic biogas, such as China and India.

The Netherlands Development Organisation, SNV, supports national programmes on domestic biogas that aim to establish commercial-viable domestic biogas sectors in which local companies market, install and service biogas plants for households. In Asia, SNV is working in Nepal, Vietnam, Bangladesh, Bhutan, Cambodia, Lao PDR, Pakistan and Indonesia, and in Africa; Rwanda, Senegal, Burkina Faso, Ethiopia, Tanzania, Uganda, Kenya, Benin and Cameroon.

In South Africa a prebuilt Biogas system is manufactured and sold. One key feature is that installation requires less skill and is quicker to install as the digester tank is premade plastic.

Society and Culture

In the 1985 Australian film *Mad Max Beyond Thunderdome* the post-apocalyptic settlement Barter town is powered by a central biogas system based upon a piggery. As well as providing electricity, methane is used to power Barter's vehicles.

"Cow Town", written in the early 1940s, discuss the travails of a city vastly built on cow manure and the hardships brought upon by the resulting methane biogas. Carter McCormick, an engineer from a town outside the city, is sent in to figure out a way to utilize this gas to help power, rather than suffocate, the city.

Energy Content of Biofuel

The Energy content of biofuel is a description of the potential energy contained in a given biofuel, measured per unit mass of that fuel, as specific energy, or per unit of volume of the fuel, as energy density. A biofuel is a fuel, produced from living organisms. Biofuels include bioethanol, an alcohol made by fermentation—often used as a gasoline additive, and biodiesel, which is usually used as a diesel additive. Specific energy is energy per unit mass, which is used to describe the energy content of a fuel, expressed in SI units as joule per kilogram (J/kg) or equivalent units. Energy density is the amount of energy stored in a fuel per unit volume, expressed in SI units as joule per litre (J/L) or equivalent units.

Energy and CO_2 output of Common Biofuels

The table below includes entries for popular substances already used for their energy, or being discussed for such use.

The second column shows specific energy, the energy content in megajoules per unit of mass in kilograms, useful in understanding the energy that can be extracted from the fuel.

The third column in the table lists energy density, the energy content per liter of volume, which is useful for understanding the space needed for storing the fuel.

The final two columns deal with the carbon footprint of the fuel. The fourth column contains the proportion of CO_2 released when the fuel is converted for energy, with respect to its starting mass, and the fifth column lists the energy produced per kilogram of CO_2 produced. As a guideline, a higher number in this column is better for the environment. But these numbers do not account for other green house gases released during burning, production, storage, or shipping. For example, methane may have hidden environmental costs that are not reflected in the table.

Fuel Type	Specific energy (MJ/kg)	Energy Density (MJ/L)	CO_2 Gas made from Fuel Used (kg/kg)	Energy per CO_2 (MJ/kg)
Solid Fuels				
Bagasse (Cane Stalks)	9.6		~+40%$(C_6H_{10}O_5)_n$+15%$(C_{26}H_{42}O_{21})_n$+15%$(C_9H_{10}O_2)_n$ 1.30	7.41
Chaff (Seed Casings)	14.6		[Please insert average composition here]	
Animal Dung/Manure	10- 15		[Please insert average composition here]	
Dried plants $(C_6H_{10}O_5)_n$	10 – 16	1.6 - 16.64	IF50%$(C_6H_{10}O_5)_n$+25%$(C_{26}H_{42}O_{21})_n$+25%$(C_{10}H_{12}O_3)_n$ 1.84	5.44-8.70
Wood fuel $(C_6H_{10}O_5)_n$	16 – 21	2.56 - 21.84	IF45%$(C_6H_{10}O_5)_n$+25%$(C_{26}H_{42}O_{21})_n$+30%$(C_{10}H_{12}O_3)_n$ 1.88	8.51-11.17
Charcoal	30		85-98% Carbon+VOC+Ash 3.63	8.27
Liquid Fuels				
Pyrolysis oil	17.5	21.35	(Assumption Of Fuel: Carbon Content = 23% w/w) 0.84	20.77
Methanol (CH_3-OH)	19.9 – 22.7	15.9	1.37	14.49-16.53
Ethanol (CH_3-CH_2-OH)	23.4 – 26.8	18.4 - 21.2	1.91	12.25-14.03
Ecalene™	28.4	22.7	75%C_2H_6O+9%C_3H_8O+7%$C_4H_{10}O$+5%$C_5H_{12}O$+4% Hx 2.03	14.02
Butanol(CH_3-$(CH_2)_3$-OH)	36	29.2	2.37	15.16
Fat	37.656	31.68	[Please insert average composition here]	
Biodiesel	37.8	33.3 – 35.7	~2.85	~13.26

Sunflower oil ($C_{18}H_{32}O_2$)	39.49	33.18	(12%($C_{16}H_{32}O_2$)+16%($C_{18}H_{34}O_2$)+71%(LA)+1%(A-LA))2.81	14.04
Castor oil ($C_{18}H_{34}O_3$)	39.5	33.21	(1%PA+1%SA+89.5 %ROA+3%OA+4.2% LA+0.3%ALA)2.67	14.80
Olive oil ($C_{18}H_{34}O_2$)	39.25 - 39.82	33 - 33.48	(15%($C_{16}H_{32}O_2$)+75%($C_{18}H_{34}O_2$)+9%(LA)+1%(A-LA))2.80	14.03
Gaseous Fuels				
Methane (CH_4)	55 – 55.7	(Liquefied) 23.0 – 23.3	(Methane leak exerts 23 × greenhouse effect of CO_2) 2.74	20.05-20.30
Hydrogen (H_2)	120 – 142	(Liquefied) 8.5 – 10.1	(Hydrogen leak slightly catalyzes ozone depletion) 0.0	
Fossil Fuels (comparison)				
Coal	29.3 – 33.5	39.85 - 74.43	(Not Counting:CO, NO_x, Sulfates & Particulates) ~3.59	~8.16-9.33
Crude Oil	41.868	28 – 31.4	(Not Counting:CO,NO_{x-} ,Sulfates & Particulates) ~3.4	~12.31
Gasoline	45 – 48.3	32 – 34.8	(Not Counting:CO,NO_{x-} ,Sulfates & Particulates) ~3.30	~13.64-14.64
Diesel	48.1	40.3	(Not Counting:CO,NO_{x-} ,Sulfates & Particulates) ~3.4	~14.15
Natural Gas	38 – 50	(Liquefied) 25.5 – 28.7	(Ethane, Propane & Butane N/C:CO,NO_x & Sulfates) ~3.00	~12.67-16.67
Ethane (CH_3-CH_3)	51.9	(Liquefied) ~24.0	2.93	17.71
Nuclear fuels (comparison)				
Uranium-235 (^{235}U)	77,000,000	(Pure)1,470,700,000	[Greater for lower ore conc.(Mining, Refining, Moving)] 0.0	~55 - ~90
Nuclear fusion (^2H-^3H)	300,000,000	(Liquefied)53,414,377.6	(Sea-Bed Hydrogen-Isotope Mining-Method Dependent) 0.0	
Fuel Cell Energy Storage (comparison)				
Direct-Methanol	4.5466	3.6	~1.37	~3.31
Proton-Exchange (R&D)	up to 5.68	up to 4.5	(IFF Fuel is recycled) 0.0	

Sodium Hydride (R&D)	up to 11.13	up to 10.24	(Bladder for Sodium Oxide Recycling) 0.0	
Battery Energy Storage (comparison)				
Lead-acid battery	0.108	~0.1	(200-600 Deep-Cycle Tolerance) 0.0	
Nickel-iron battery	0.0487 - 0.1127	0.0658 - 0.1772	(<40y Life)(2k-3k Cycle Tolerance IF no Memory effect) 0.0	
Nickel-cadmium battery	0.162 - 0.288	~0.24	(1k-1.5k Cycle Tolerance IF no Memory effect) 0.0	
Nickel metal hydride	0.22 - 0.324	0.36	(300-500 Cycle Tolerance IF no Memory effect) 0.0	
Super iron battery	0.33	(1.5 * NiMH) 0.54	(~300 Deep-Cycle Tolerance) 0.0	
Zinc-air battery	0.396 - 0.72	0.5924 - 0.8442	(Recyclable by Smelting & Remixing, not Recharging) 0.0	
Lithium ion battery	0.54 - 0.72	0.9 - 1.9	(3-5 y Life) (500-1k Deep-Cycle Tolerance) 0.0	
Lithium-Ion-Polymer	0.65 - 0.87	(1.2 * Li-Ion)1.08 - 2.28	(3-5 y Life) (300-500 Deep-Cycle Tolerance) 0.0	
Lithium iron phosphate battery				
DURACELL Zinc-Air	1.0584 - 1.5912	5.148 - 6.3216	(1-3 y Shelf-life) (Recyclable not Rechargeable) 0.0	
Aluminium battery	1.8 - 4.788	7.56	(10-30 y Life) (3k+ Deep-Cycle Tolerance) 0.0	
PolyPlusBC Li-Air-cell	3.6 - 32.4	3.6 - 17.64	(May be Rechargeable) (Might leak sulfates) 0.0	

Yields of Common Crops Associated with Biofuels Production

This table below needs citations and explanation of methodology!

Crop	**Oil (kg/ha)**	**Oil (L/ha)**	**Oil (lb/acre)**	**Oil (US gal/acre)**	**Oil per seeds (kg/100 kg)**	**Melting Range (°C)**			**Iodine number**	**Cetane number**
						Oil / Fat	**Methyl Ester**	**Ethyl Ester**		
Groundnut					(Kernel)42					

Copra					62					
Tallow						35 - 42	16	12	40 - 60	75
Lard						32 - 36	14	10	60 - 70	65
Corn (maize)	145	172	129	18		-5	-10	-12	115 - 124	53
Cashew nut	148	176	132	19						
Oats	183	217	163	23						
Lupine	195	232	175	25						
Kenaf	230	273	205	29						
Calen-dula	256	305	229	33						
Cotton	273	325	244	35	(Seed)13	-1 - 0	-5	-8	100 - 115	55
Hemp	305	363	272	39						
Soybean	375	446	335	48	14	-16 - -12	-10	-12	125 - 140	53
Coffee	386	459	345	49						
Linseed (flax)	402	478	359	51		-24			178	
Hazel-nuts	405	482	362	51						
Euphor-bia	440	524	393	56						
Pump-kin seed	449	534	401	57						
Corian-der	450	536	402	57						
Mustard seed	481	572	430	61	35					
Cameli-na	490	583	438	62						
Sesame	585	696	522	74	50					
Safflow-er	655	779	585	83						
Rice	696	828	622	88						
Tung oil tree	790	940	705	100		-2.5			168	
Sun-flowers	800	952	714	102	32	-18 - -17	-12	-14	125 - 135	52

Cocoa (cacao)	863	1,026	771	110						
Peanuts	890	1,059	795	113		3			93	
Opium poppy	978	1,163	873	124						
Rape-seed	1,000	1,190	893	127	37	-10 - 5	-10 - 0	-12 - -2	97 - 115	55 - 58
Olives	1,019	1,212	910	129		-12 - -6	-6	-8	77 - 94	60
Castor beans	1,188	1,413	1,061	151	(Seed)50	-18			85	
Pecan nuts	1,505	1,791	1,344	191						
Jojoba	1,528	1,818	1,365	194						
Jatro-pha	1,590	1,892	1,420	202						
Maca-damia nuts	1,887	2,246	1,685	240						
Brazil nuts	2,010	2,392	1,795	255						
Avocado	2,217	2,638	1,980	282						
Coconut	2,260	2,689	2,018	287		20 - 25	-9	-6	8 - 10	70
Chinese Tallow		4,700		500						
Oil palm	5,000	5,950	4,465	635	20-(Ker-nal)36	20 - 40	-8 - 21	-8 - 18	12 - 95	65 - 85
Algae		95,000		10,000						
Crop	**Oil (kg/ha)**	**Oil (L/ha)**	**Oil (lb/acre)**	**Oil (US gal/acre)**	**Oil per seeds (kg/100 kg)**	**Melting Range (°C)**			**Iodine number**	**Ce-tane number**
						Oil / Fat	**Meth-yl Ester**	**Eth-yl Es-ter**		

Sustainable Biofuel

Biofuels, in the form of liquid fuels derived from plant materials, are entering the market, driven by factors such as oil price spikes and the need for increased energy security.

However, many of the biofuels that are currently being supplied have been criticised for their adverse impacts on the natural environment, food security, and land use.

The challenge is to support biofuel development, including the development of new cellulosic technologies, with responsible policies and economic instruments to help ensure that biofuel commercialization is sustainable. Responsible commercialization of biofuels represents an opportunity to enhance sustainable economic prospects in Africa, Latin America and Asia.

Biofuels have a limited ability to replace fossil fuels and should not be regarded as a 'silver bullet' to deal with transport emissions. However, they offer the prospect of increased market competition and oil price moderation. A healthy supply of alternative energy sources will help to combat gasoline price spikes and reduce dependency on fossil fuels, especially in the transport sector. Using transportation fuels more efficiently is also an integral part of a sustainable transport strategy.

Biofuel Options

Biofuel development and use is a complex issue because there are many biofuel options which are available. Biofuels, such as ethanol and biodiesel, are currently produced from the products of conventional food crops such as the starch, sugar and oil feedstocks from crops that include wheat, maize, sugar cane, palm oil and oilseed rape. Some researchers fear that a major switch to biofuels from such crops would create a direct competition with their use for food and animal feed, and claim that in some parts of the world the economic consequences are already visible, other researchers look at the land available and the enormous areas of idle and abandoned land and claim that there is room for a large proportion of biofuel also from conventional crops.

Second generation biofuels are now being produced from a much broader range of feedstocks including the cellulose in dedicated energy crops (perennial grasses such as switchgrass and Miscanthus giganteus), forestry materials, the co-products from food production, and domestic vegetable waste. Advances in the conversion processes will improve the sustainability of biofuels, through better efficiencies and reduced environmental impact of producing biofuels, from both existing food crops and from cellulosic sources.

In 2007, Ronald Oxburgh suggested in *The Courier-Mail* that production of biofuels could be either responsible or irresponsible and had several trade-offs: "Produced responsibly they are a sustainable energy source that need not divert any land from growing food nor damage the environment; they can also help solve the problems of the waste generated by Western society; and they can create jobs for the poor where previously were none. Produced irresponsibly, they at best offer no climate benefit and, at worst, have detrimental social and environmental consequences. In other words, biofuels are pretty much like any other product. In 2008 the Nobel prize-winning chemist

Paul J. Crutzen published findings that the release of nitrous oxide (N_2O) emissions in the production of biofuels means that they contribute more to global warming than the fossil fuels they replace.

According to the Rocky Mountain Institute, sound biofuel production practices would not hamper food and fibre production, nor cause water or environmental problems, and would enhance soil fertility. The selection of land on which to grow the feedstocks is a critical component of the ability of biofuels to deliver sustainable solutions. A key consideration is the minimisation of biofuel competition for prime cropland.

Plants Used as Sustainable Biofuel

Sugarcane in Brazil

Brazil's production of ethanol fuel from sugarcane dates back to the 1970s, as a governmental response to the 1973 oil crisis. Brazil is considered the biofuel industry leader and the world's first sustainable biofuels economy. In 2010 the U.S. Environmental Protection Agency designated Brazilian sugarcane ethanol as an advanced biofuel due to EPA's estimated 61% reduction of total life cycle greenhouse gas emissions, including direct indirect land use change emissions. Brazil sugarcane ethanol fuel program success and sustainability is based on the most efficient agricultural technology for sugarcane cultivation in the world, uses modern equipment and cheap sugar cane as feedstock, the residual cane-waste (bagasse) is used to process heat and power, which results in a very competitive price and also in a high energy balance (output energy/input energy), which varies from 8.3 for average conditions to 10.2 for best practice production.

Sugarcane (*Saccharum officinarum*) plantation ready for harvest, Ituverava, São Paulo State, Brazil.

A report commissioned by the United Nations, based on a detailed review of published research up to mid-2009 as well as the input of independent experts world-wide, found that ethanol from sugar cane as produced in Brazil "*in some circumstances does better than just "zero emission". If grown and processed correctly, it has negative emission, pulling CO2 out of the atmosphere, rather than adding it.* In contrast, the report found that U.S. use of maize for biofuel is less efficient, as sugarcane can lead to emissions reductions of between 70% and well over 100% when substituted for gasoline. Several other studies have shown that sugarcane-based ethanol reduces greenhouse gases by 86 to 90% if there is no significant land use change.

Mechanized harvesting of sugarcane, Piracicaba, São Paulo, Brazil.

In another study commissioned by the Dutch government in 2006 to evaluate the sustainability of Brazilian bioethanol concluded that there is sufficient water to supply all foreseeable long-term water requirements for sugarcane and ethanol production. This evaluation also found that consumption of agrochemicals for sugar cane production is lower than in citric, corn, coffee and soybean cropping. The study found that development of resistant sugar cane varieties is a crucial aspect of disease and pest control and is one of the primary objectives of Brazil's cane genetic improvement programs. Disease control is one of the main reasons for the replacement of a commercial variety of sugar cane.

Cosan's Costa Pinto sugar cane mill and ethanol distillery plant at Piracicaba, São Paulo, Brazil.

Another concern is the fact that sugarcane fields are traditionally burned just before harvest to avoid harm to the workers, by removing the sharp leaves and killing snakes and other harmful animals, and also to fertilize the fields with ash. Mechanization will reduce pollution from burning fields and has higher productivity than people, and due to mechanization the number of temporary workers in the sugarcane plantations has already declined. By the 2008 harvest season, around 47% of the cane was collected with harvesting machines.

Regarding the negative impacts of the potential direct and indirect effect of land use changes on carbon emissions, the study commissioned by the Dutch government concluded that "it is very difficult to determine the indirect effects of further land use for sugar cane production (i.e. sugar cane replacing another crop like soy or citrus crops, which in turn causes additional soy plantations replacing pastures, which in turn may cause deforestation), and also not logical to attribute all these soil carbon losses to sugar cane". The Brazilian agency Embrapa estimates that there is enough agricultural land available to increase at least 30 times the existing sugarcane plantation without endangering sensible ecosystems or taking land destined for food crops. Most future growth is expected to take place on abandoned pasture lands, as it has been the historical trend in São Paulo state. Also, productivity is expected to improve even further based on current biotechnology research, genetic improvement, and better agronomic practices, thus contributing to reduce land demand for future sugarcane cultures.

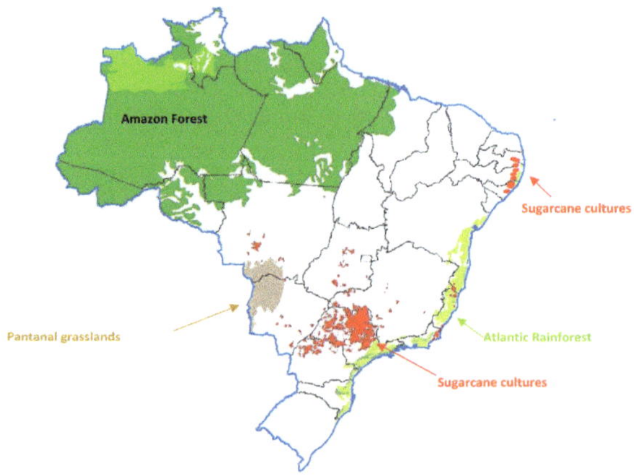

Location of environmentally valuable areas with respect to sugarcane plantations. São Paulo, located in the Southeast Region of Brazil, concentrates two-thirds of sugarcane cultures.

Another concern is the risk of clearing rain forests and other environmentally valuable land for sugarcane production, such as the Amazonia, the Pantanal or the Cerrado. Embrapa has rebutted this concern explaining that 99.7% of sugarcane plantations are located at least 2,000 km from the Amazonia, and expansion during the last 25 years took place in the Center-South region, also far away from the Amazonia, the Pantanal or the Atlantic forest. In São Paulo state growth took place in abandoned pasture lands. The impact assessment commissioned by the Dutch government supported this argument.

In order to guarantee a sustainable development of ethanol production, in September 2009 the government issued by decree a countrywide agroecological land use zoning to restrict sugarcane growth in or near environmentally sensitive areas. According to the new criteria, 92.5% of the Brazilian territory is not suitable for sugarcane plantation. The government considers that the suitable areas are more than enough to meet the future demand for ethanol and sugar in the domestic and international markets foreseen for the next decades.

Regarding the food vs fuel issue, a World Bank research report published on July 2008 found that "*Brazil's sugar-based ethanol did not push food prices appreciably higher*". This research paper also concluded that Brazil's sugar cane–based ethanol has not raised sugar prices significantly. An economic assessment report also published in July 2008 by the OECD agrees with the World Bank report regarding the negative effects of subsidies and trade restrictions, but found that the impact of biofuels on food prices are much smaller. A study by the Brazilian research unit of the Fundação Getúlio Vargas regarding the effects of biofuels on grain prices concluded that the major driver behind the 2007-2008 rise in food prices was speculative activity on futures markets under conditions of increased demand in a market with low grain stocks. The study also concluded that there is no correlation between Brazilian sugarcane cultivated area and average grain prices, as on the contrary, the spread of sugarcane was accompanied by rapid growth of grain crops in the country.

Jatropha

India and Africa

Jatropha gossipifolia in Hyderabad, India.

Crops like Jatropha, used for biodiesel, can thrive on marginal agricultural land where many trees and crops won't grow, or would produce only slow growth yields. Jatropha cultivation provides benefits for local communities:

Cultivation and fruit picking by hand is labour-intensive and needs around one person per hectare. In parts of rural India and Africa this provides much-needed jobs - about 200,000 people worldwide now find employment through jatropha. Moreover, villagers often find that they can grow other crops in the shade of the trees. Their communities will avoid importing expensive diesel and there will be some for export too.

Cambodia

Cambodia has no proven fossil fuel reserves, and is almost completely dependent on imported diesel fuel for electricity production. Consequently, Cambodians face an insecure supply and pay some of the highest energy prices in the world. The impacts of this are widespread and may hinder economic development.

Biofuels may provide a substitute for diesel fuel that can be manufactured locally for a lower price, independent of the international oil price. The local production and use of biofuel also offers other benefits such as improved energy security, rural development opportunities and environmental benefits. The Jatropha curcas species appears to be a particularly suitable source of biofuel as it already grows commonly in Cambodia. Local sustainable production of biofuel in Cambodia, based on the Jatropha or other sources, offers good potential benefits for the investors, the economy, rural communities and the environment.

Mexico

Jatropha is native to Mexico and Central America and was likely transported to India and Africa in the 1500s by Portuguese sailors convinced it had medicinal uses. In 2008, recognizing the need to diversify its sources of energy and reduce emissions, Mexico passed a law to push developing biofuels that don't threaten food security and the agriculture ministry has since identified some 2.6 million hectares (6.4 million acres) of land with a high potential to produce jatropha. The Yucatán Peninsula, for instance, in addition to being a corn producing region, also contains abandoned sisal plantations, where the growing of Jatropha for biodiesel production would not displace food.

On April 1, 2011 Interjet completed the first Mexican aviation biofuels test flight on an Airbus A320. The fuel was a 70:30 traditional jet fuel biojet blend produced from Jatropha oil provided by three Mexican producers, Global Energías Renovables (a wholly owned subsidiary of U.S.-based Global Clean Energy Holdings, Bencafser S.A. and Energy JH S.A. Honeywell's UOP processed the oil into Bio-SPK (Synthetic Paraffinic Kerosene) . Global Energías Renovables operates the largest Jatropha farm in the Americas.

On August 1, 2011 Aeromexico, Boeing, and the Mexican Government participated in the first biojet powered transcontinental flight in aviation history. The flight from Mexico City to Madrid used a blend of 70 percent traditional fuel and 30 percent biofuel (aviation biofuel). The biojet was produced entirely from Jatropha oil.

Pongamia Pinnata in Australia and India

Pongamia pinnata is a legume native to Australia, India, Florida (USA) and most tropical regions, and is now being invested in as an alternative to Jatropha for areas such as Northern Australia, where Jatropha is classed as a noxious weed. Commonly known as simply 'Pongamia', this tree is currently being commercialised in Australia by Pacific Renewable Energy, for use as a Diesel replacement for running in modified Diesel engines or for conversion to Biodiesel using 1st or 2nd Generation Biodiesel techniques, for running in unmodified Diesel engines.

Pongamia pinnata seeds in Brisbane, Australia.

Sweet Sorghum in India

Sweet sorghum overcomes many of the shortcomings of other biofuel crops. With sweet sorghum, only the stalks are used for biofuel production, while the grain is saved for food or livestock feed. It is not in high demand in the global food market, and thus has little impact on food prices and food security. Sweet sorghum is grown on already-farmed drylands that are low in carbon storage capacity, so concerns about the clearing of rainforest do not apply. Sweet sorghum is easier and cheaper to grow than other biofuel crops in India and does not require irrigation, an important consideration in dry areas. Some of the Indian sweet sorghum varieties are now grown in Uganda for ethanol production.

A study by researchers at the International Crops Research Institute for the Semi-Arid Tropics (ICRISAT) found that growing sweet sorghum instead of grain sorghum could increase farmers incomes by US$40 per hectare per crop because it can provide food, feed and fuel. With grain sorghum currently grown on over 11 million hectares (ha) in Asia and on 23.4 million ha in Africa, a switch to sweet sorghum could have a considerable economic impact.

International Collaboration on Sustainable Biofuels

Roundtable on Sustainable Biofuels

Public attitudes and the actions of key stakeholders can play a crucial role in realising the potential of sustainable biofuels. Informed discussion and dialogue, based both on scientific research and an understanding of public and stakeholder views, is important.

The Roundtable on Sustainable Biofuels is an international initiative which brings together farmers, companies, governments, non-governmental organizations, and scientists who are interested in the sustainability of biofuels production and distribution. During 2008, the Roundtable used meetings, teleconferences, and online discussions to develop a series of principles and criteria for sustainable biofuels production.

In 2008, the Roundtable for Sustainable Biofuels released its proposed standards for sustainable biofuels. This includes 12 principles:

1. "Biofuel production shall follow international treaties and national laws regarding such things as air quality, water resources, agricultural practices, labor conditions, and more.
2. Biofuels projects shall be designed and operated in participatory processes that involve all relevant stakeholders in planning and monitoring.
3. Biofuels shall significantly reduce greenhouse gas emissions as compared to fossil fuels. The principle seeks to establish a standard methodology for comparing greenhouse gases (GHG) benefits.
4. Biofuel production shall not violate human rights or labor rights, and shall ensure decent work and the well-being of workers.
5. Biofuel production shall contribute to the social and economic development of local, rural and indigenous peoples and communities.
6. Biofuel production shall not impair food security.
7. Biofuel production shall avoid negative impacts on biodiversity, ecosystems and areas of high conservation value.
8. Biofuel production shall promote practices that improve soil health and minimize degradation.
9. Surface and groundwater use will be optimized and contamination or depletion of water resources minimized.
10. Air pollution shall be minimized along the supply chain.
11. Biofuels shall be produced in the most cost-effective way, with a commitment to improve production efficiency and social and environmental performance in all stages of the biofuel value chain.
12. Biofuel production shall not violate land rights".

In April 2011, the Roundtable on Sustainable Biofuels launched a set of comprehensive sustainability criteria - the "RSB Certification System." Biofuels producers that meet to these criteria are able to show buyers and regulators that their product has been obtained without harming the environment or violating human rights.

Sustainable Biofuels Consensus

The Sustainable Biofuels Consensus is an international initiative which calls upon governments, the private sector, and other stakeholders to take decisive action to ensure the sustainable trade, production, and use of biofuels. In this way biofuels may play a key role in energy sector transformation, climate stabilization, and resulting worldwide revitalisation of rural areas.

The Sustainable Biofuels Consensus envisions a "landscape that provides food, fodder, fiber, and energy, which offers opportunities for rural development; that diversifies energy supply, restores ecosystems, protects biodiversity, and sequesters carbon".

Better Sugarcane Initiative / Bonsucro

In 2008, a multi-stakeholder process was initiated by the World Wildlife Fund and the International Finance Corporation, the private development arm of the World Bank, bringing together industry, supply chain intermediaries, end-users, farmers and civil society organisations to develop standards for certifying the derivative products of sugar cane, one of which is ethanol fuel.

The Bonsucro standard is based around a definition of sustainability which is founded on five principles:

1. Obey the law
2. Respect human rights and labour standards
3. Manage input, production and processing efficiencies to enhance sustainability
4. Actively manage biodiversity and ecosystem services
5. Continuously improve key areas of the business

Biofuel producers that wish to sell products marked with the Bonsucro standard must both ensure that they product to the Production Standard, and that their downstream buyers meet the Chain of Custody Standard. In addition, if they wish to sell to the European market and count against the EU Renewable Energy Directive, then they must adhere to the Bonsucro EU standard, which includes specific greenhouse gas calculations following European Commission calculation guidelines.

Sustainability Standards

Several countries and regions have introduced policies or adopted standards to promote sustainable biofuels production and use, most prominently the European Union and the United States. The 2009 EU Renewable Energy Directive, which requires 10 percent of transportation energy from renewable energy by 2020, is the most comprehensive mandatory sustainability standard in place as of 2010. The Directive requires that the lifecycle greenhouse gas emissions of biofuels consumed be at least 50 percent less than the equivalent emissions from gasoline or diesel by 2017 (and 35 percent less starting in 2011). Also, the feedstocks for biofuels "should not be harvested from lands with high biodiversity value, from carbon-rich or forested land, or from wetlands".

As with the EU, the U.S. Renewable Fuel Standard (RFS) and the California Low Carbon Fuel Standard (LCFS) both require specific levels of lifecycle greenhouse gas reductions compared to equivalent fossil fuel consumption. The RFS requires that at least half of the biofuels production mandated by 2022 should reduce lifecycle emissions by 50 percent. The LCFS is a performance standard that calls for a minimum of 10 percent emissions reduction per unit of transport energy by 2020. Both the U.S. and California standards currently address only greenhouse gas emissions, but California plans to "expand its policy to address other sustainability issues associated with liquid biofuels in the future".

In 2009, Brazil also adopted new sustainability policies for sugarcane ethanol, including "zoning regulation of sugarcane expansion and social protocols".

Oil Price Moderation

Biofuels offer the prospect of real market competition and oil price moderation. According to the Wall Street Journal, crude oil would be trading 15 per cent higher and gasoline would be as much as 25 per cent more expensive, if it were not for biofuels. A healthy supply of alternative energy sources will help to combat gasoline price spikes.

Sustainable Transport

Biofuels have a limited ability to replace fossil fuels and should not be regarded as a 'silver bullet' to deal with transport emissions. Biofuels on their own cannot deliver a sustainable transport system and so must be developed as part of an integrated approach, which promotes other renewable energy options and energy efficiency, as well as reducing the overall energy demand and need for transport. Consideration needs to be given to the development of hybrid and fuel cell vehicles, public transport, and better town and rural planning.

In December 2008 an Air New Zealand jet completed the world's first commercial aviation test flight partially using jatropha-based fuel. More than a dozen performance tests were undertaken in the two-hour test flight which departed from Auckland International Airport. A biofuel blend of 50:50 jatropha and Jet A1 fuel was used to power one of the Boeing 747-400's Rolls-Royce RB211 engines. Air New Zealand set several criteria for its jatropha, requiring that "the land it came from was neither forest nor virgin grassland in the previous 20 years, that the soil and climate it came from is not suitable for the majority of food crops and that the farms are rain fed and not mechanically irrigated". The company has also set general sustainability criteria, saying that such biofuels must not compete with food resources, that they must be as good as traditional jet fuels, and that they should be cost competitive.

In January 2009, Continental Airlines used a sustainable biofuel to power a commercial aircraft for the first time in North America. This demonstration flight marks the first sustainable biofuel demonstration flight by a commercial carrier using a twin-engined aircraft, a Boeing 737-800, powered by CFM International CFM56-7B engines. The biofuel blend included components derived from algae and jatropha plants. The algae oil was provided by Sapphire Energy, and the jatropha oil by Terasol Energy.

In March 2011, Yale University research showed significant potential for sustainable aviation fuel based on jatropha-curcas. According to the research, if cultivated properly, "jatropha can deliver many benefits in Latin America and greenhouse gas reductions of up to 60 percent when compared to petroleum-based jet fuel". Actual farming conditions in Latin America were assessed using sustainability criteria developed by the Roundtable on Sustainable Biofuels. Unlike previous research, which used theoretical inputs, the Yale team conducted many interviews with jatropha farmers and used "field measurements to develop the first comprehensive sustainability analysis of actual projects".

As of June 2011, revised international aviation fuel standards officially allow commercial airlines to blend conventional jet fuel with up to 50 percent biofuels. The renewable fuels "can be blended with conventional commercial and military jet fuel through requirements in the newly issued edition of ASTM D7566, Specification for Aviation Turbine Fuel Containing Synthesized Hydrocarbons".

In December 2011, the FAA awarded $7.7 million to eight companies to advance the development of commercial aviation biofuels, with a special focus on alcohol to jet fuel. The FAA is assisting in the development of a sustainable fuel (from alcohols, sugars, biomass, and organic matter such as pyrolysis oils) that can be "dropped in" to aircraft without changing current practices and infrastructure. The research will test how the new fuels affect engine durability and quality control standards.

GreenSky London, a biofuels plant under construction in 2014, will take in some 500,000 tonnes of municipal rubbish and change the organic component into 60,000 tonnes of jet fuel, and 40 megawatts of power. By the end of 2015, all British Airways flights from London City Airport will be fuelled by waste and rubbish discarded by London residents.

Second-generation Biofuels

Second generation biofuels, also known as advanced biofuels, are fuels that can be manufactured from various types of biomass. Biomass is a wide-ranging term meaning any source of organic carbon that is renewed rapidly as part of the carbon cycle. Biomass is derived from plant materials but can also include animal materials.

First generation biofuels are made from the sugars and vegetable oils found in arable crops, which can be easily extracted using conventional technology. In comparison, second generation biofuels are made from lignocellulosic biomass or woody crops, agricultural residues or waste, which makes it harder to extract the required fuel.

Introduction

Second generation biofuel technologies have been developed because first generation biofuels manufacture has important limitations. First generation biofuel processes are useful but limited in most cases: there is a threshold above which they cannot produce enough biofuel without threatening food supplies and biodiversity. Many first generation biofuels depend on subsidies and are not cost competitive with existing fossil fuels such as oil, and some of them produce only limited greenhouse gas emissions savings. When taking emissions from production and transport into account, life-cycle assessment from first generation biofuels frequently approach those of traditional fossil fuels.

Second generation biofuels can help solve these problems and can supply a larger proportion of global fuel supply sustainably, affordably, and with greater environmental benefits.

First generation bioethanol is produced by fermenting plant-derived sugars to ethanol, using a similar process to that used in beer and wine-making. This requires the use of 'food' crops, such as sugar cane, corn, wheat, and sugar beet. These crops are required for food, so, if too much biofuel is made from them, food prices could rise and shortages might be experienced in some countries. Corn, wheat, and sugar beet can also require high agricultural inputs in the form of fertilizers, which limit the greenhouse gas reductions that can be achieved. Biodiesel produced by transesterification from rapeseed oil, palm oil, or other plant oils is also considered a first generation biofuel.

The goal of second generation biofuel processes is to extend the amount of biofuel that can be produced sustainably by using biomass consisting of the residual non-food parts of current crops, such as stems, leaves and husks that are left behind once the food crop has been extracted, as well as other crops that are not used for food purposes (non-food crops), such as switchgrass, grass, jatropha, whole crop maize, miscanthus and cereals that bear little grain, and also industry waste such as woodchips, skins and pulp from fruit pressing, etc.

The problem that second generation biofuel processes are addressing is to extract useful feedstocks from this woody or fibrous biomass, where the useful sugars are locked in by lignin, hemicellulose and cellulose. All plants contain lignin, hemicellulose and cellulose. These are complex carbohydrates (molecules based on sugar). Lignocellulosic ethanol is made by freeing the sugar molecules from cellulose using enzymes, steam heating, or other pre-treatments. These sugars can then be fermented to produce ethanol in the same way as first generation bioethanol production. The by-product of this process is lignin. Lignin can be burned as a carbon neutral fuel to produce heat and power for the processing plant and possibly for surrounding homes and businesses.

Thermochemical processes (liquefaction) in hydrothermal media can produce liquid oily products from a wide range of feedstock that has a potential to replace or augment fuels. However, these liquid products fall short of diesel or biodiesel standards. Upgrading liquefaction products through one or many physical or chemical processes may improve properties for use as fuel.

Second Generation Technology

The following subsections describe the main second generation routes currently under development.

Thermochemical Routes

Carbon-based materials can be heated at high temperatures in the absence (pyrolysis) or presence of oxygen, air and/or steam (gasification).

These thermochemical processes both yield a combustible gas and a solid char. The gas can be fermented or chemically synthesised into a range of fuels, including ethanol, synthetic diesel or jet fuel.

There are also lower temperature processes in the region of 150-374 °C, that produce sugars by decomposing the biomass in water with or without additives.

Gasification

Gasification technologies are well established for conventional feedstocks such as coal and crude oil. Second generation gasification technologies include gasification of forest and agricultural residues, waste wood, energy crops and black liquor. Output is normally syngas for further synthesis to e.g. Fischer-Tropsch products including diesel fuel, bio-methanol, BioDME (dimethyl ether), gasoline via catalytic conversion of dimethyl ether, or biomethane (synthetic natural gas). Syngas can also be used in heat production and for generation of mechanical and electrical power via gas motors or gas turbines.

Pyrolysis

Pyrolysis is a well established technique for decomposition of organic material at elevated temperatures in the absence of oxygen. In second generation biofuels applications forest and agricultural residues, wood waste and energy crops can be used as feedstock to produce e.g. bio-oil for fuel oil applications. Bio-oil typically requires significant additional treatment to render it suitable as a refinery feedstock to replace crude oil.

Torrefaction

Torrefaction is a form of pyrolysis at temperatures typically ranging between 200-320 °C. Feedstocks and output are the same as for pyrolysis.

Biochemical Routes

Chemical and biological processes that are currently used in other applications are being adapted for second generation biofuels. Biochemical processes typically employ pre-treatment to accelerate the hydrolysis process, which separates out the lignin, hemicellulose and cellulose. Once these ingredients are separated, the cellulose fractions can be fermented into alcohols.

Feedstocks are energy crops, agricultural and forest residues, food industry and municipal biowaste and other biomass containing sugars. Products include alcohols (such as ethanol and butanol) and other hydrocarbons for transportation use.

Types of Biofuel

The following second generation biofuels are under development, although most or all of these biofuels are synthesized from intermediary products such as syngas using methods that are identical in processes involving conventional feedstocks, first generation and second generation biofuels. The distinguishing feature is the technology involved in producing the intermediary product, rather than the ultimate off-take.

A process producing liquid fuels from gas (normally syngas) is called a Gas-to-Liquid (GtL) process. When biomass is the source of the gas production the process is also referred to as Biomass-To-Liquids (BTL).

From Syngas Using Catalysis

- Biomethanol can be used in methanol motors or blended with petrol up to 10-20% without any infrastructure changes.
- BioDME can be produced from Biomethanol using catalytic dehydration or it can be produced directly from syngas using direct DME synthesis. DME can be used in the compression ignition engine.
- Bio-derived gasoline can be produced from DME via high-pressure catalytic condensation reaction. Bio-derived gasoline is chemically indistinguishable from petroleum-derived gasoline and thus can be blended into the U.S. gasoline pool.
- can be used in fuel cells to produce electricity.
- Mixed Alcohols (i.e., mixture of mostly ethanol, propanol, and butanol, with some pentanol, hexanol, heptanol, and octanol). Mixed alcohols are produced from syngas with several classes of catalysts. Some have employed catalysts similar to those used for methanol. Molybdenum sulfide catalysts were discovered at Dow Chemical and have received considerable attention. Addition of cobalt sulfide to the catalyst formulation was shown to enhance performance. Molyb-

denum sulfide catalysts have been well studied but have yet to find widespread use. These catalysts have been a focus of efforts at the U.S. Department of Energy's Biomass Program in the Thermochemical Platform. Noble metal catalysts have also been shown to produce mixed alcohols. Most R&D in this area is concentrated in producing mostly ethanol. However, some fuels are marketed as mixed alcohols Mixed alcohols are superior to pure methanol or ethanol, in that the higher alcohols have higher energy content. Also, when blending, the higher alcohols increase compatibility of gasoline and ethanol, which increases water tolerance and decreases evaporative emissions. In addition, higher alcohols have also lower heat of vaporization than ethanol, which is important for cold starts.

- Biomethane (or Bio-SNG) via the Sabatier reaction

From Syngas Using Fischer-Tropsch

The Fischer-Tropsch (FT) process is a Gas-to-Liquid (GtL) process. When biomass is the source of the gas production the process is also referred to as Biomass-To-Liquids (BTL). A disadvantage of this process is the high energy investment for the FT synthesis and consequently, the process is not yet economic.

- FT diesel can be mixed with fossil diesel at any percentage without need for infrastructure change and moreover, synthetic kerosene can be produced

Biocatalysis

- Biohydrogen might be accomplished with some organisms that produce hydrogen directly under certain conditions. Biohydrogen can be used in fuel cells to produce electricity.
- Butanol and Isobutanol via recombinant pathways expressed in hosts such as E. coli and yeast, butanol and isobutanol may be significant products of fermentation using glucose as a carbon and energy source.
- (2,5-Dimethylfuran). Recent advances in producing DMF from fructose and glucose using catalytic biomass-to-liquid process have increased its attractiveness.

Other Processes

- HTU (Hydro Thermal Upgrading) diesel is produced from wet biomass. It can be mixed with fossil diesel in any percentage without need for infrastructure.
- A new biofuel was developed by the University of Georgia from woodchips. The oil is extracted and then added to unmodified diesel engines. Either new plants

are used or planted to replace the old plants. The charcoal byproduct is put back into the soil as a fertilizer. According to the director Tom Adams since carbon is put back into the soil, this biofuel can actually be carbon negative not just carbon neutral. Carbon negative decreases carbon dioxide in the air reversing the greenhouse effect not just reducing it.

Feedstocks

Second generation biofuel feedstocks include cereal and sugar crops, specifically grown energy crops, agricultural and municipal wastes, cultivated and waste oils, and algae. Land use, existing biomass industries and relevant conversion technologies must be considered when evaluating suitability of developing biomass as feedstock for energy.

Energy Crops

Plants are made from lignin, hemicellulose and cellulose; second generation technology uses one, two or all of these components. Common lignocellulosic energy crops include wheat straw, Miscanthus, short rotation coppice poplar and willow. However, each offers different opportunities and no one crop can be considered 'best' or 'worst'.

Municipal Solid Waste

Municipal Solid Waste comprises a very large range of materials, and total waste arisings are increasing. In the UK, recycling initiatives decrease the proportion of waste going straight for disposal, and the level of recycling is increasing each year. However, there remains significant opportunities to convert this waste to fuel via gasification or pyrolysis.

Green Waste

Green waste such as forest residues or garden or park waste may be used to produce biofuel via different routes. Examples include Biogas captured from biodegradable green waste, and gasification or hydrolysis to syngas for further processing to biofuels via catalytic processes.

Black Liquor

Black liquor, the spent cooking liquor from the kraft process that contains concentrated lignin and hemicellulose, may be gasified with very high conversion efficiency and greenhouse gas reduction potential to produce syngas for further synthesis to e.g. biomethanol or BioDME.

Greenhouse Gas Emissions

Lignocellulosic biofuels reduces greenhouse gas emissions with 60-90% when compared with fossil petroleum (Börjesson.P. et al. 2013. Dagens och framtidens hållbara biodrivmedel), which is on par with the better of current biofuels of the first generation, where typical best values currently is 60-80%. In 2010, average savings of biofuels used within EU was 60% (Hamelinck.C. et al. 2013 Renewable energy progress and biofuels sustainability, Report for the European Commission). In 2013, 70% of the biofuels used in Sweden reduced emissions with 66% or higher. (Energimyndigheten 2014. Hållbara biodrivmedel och flytande biobränslen 2013).

Commercial Development

An operating lignocellulosic ethanol production plant is located in Canada, run by Iogen Corporation. The demonstration-scale plant produces around 700,000 litres of bioethanol each year. A commercial plant is under construction. Many further lignocellulosic ethanol plants have been proposed in North America and around the world.

The Swedish specialty cellulose mill Domsjö Fabriker in Örnsköldsvik, Sweden develops a biorefinery using Chemrec's black liquor gasification technology. When commissioned in 2015 the biorefinery will produce 140,000 tons of biomethanol or 100,000 tons of BioDME per year, replacing 2% of Sweden's imports of diesel fuel for transportation purposes. In May 2012 it was revealed that Domsjö pulled out of the project, effectively killing the effort.

In the UK, companies like INEOS Bio and British Airways are developing advanced biofuel refineries, which are due to be built by 2013 and 2014 respectively. Under favourable economic conditions and strong improvements in policy support, NNFCC projections suggest advanced biofuels could meet up to 4.3 per cent of the UK's transport fuel by 2020 and save 3.2 million tonnes of CO_2 each year, equivalent to taking nearly a million cars off the road.

Helsinki, Finland, 1 February 2012 - UPM is to invest in a biorefinery producing biofuels from crude tall oil in Lappeenranta, Finland. The industrial scale investment is the first of its kind globally. The biorefinery will produce annually approximately 100,000 tonnes of advanced second generation biodiesel for transport. Construction of the biorefinery will begin in the summer of 2012 at UPM's Kaukas mill site and be completed in 2014. UPM's total investment will amount to approximately EUR 150 million.

Calgary, Alberta, 30 April 2012 – Iogen Energy Corporation has agreed to a new plan with its joint owners Royal Dutch Shell and Iogen Corporation to refocus its strategy and activities. Shell continues to explore multiple pathways to find a commercial solution for the production of advanced biofuels on an industrial scale, but the company will NOT pursue the project it has had under development to build a larger scale cellulosic ethanol facility in southern Manitoba.

"Drop-in" Biofuels

So-called "drop-in" biofuels can be defined as "liquid bio-hydrocarbons that are functionally equivalent to petroleum fuels and are fully compatible with existing petroleum infrastructure".

There is considerable interest in developing advanced biofuels that can be readily integrated in the existing petroleum fuel infrastructure - i.e. “dropped-in” - particularly by sectors such as aviation, where there are no real alternatives to sustainably produced biofuels for low carbon emitting fuel sources. Drop-in biofuels by definition should be fully fungible and compatible with the large existing “petroleum-based” infrastructure.

According to a recent report published by the IEA Bioenergy Task 39, entitled “The potential and challenges of drop-in biofuels”, there are several ways to produce drop-in biofuels that are functionally equivalent to petroleum-derived transportation fuel blendstocks. These are discussed within three major sections of the full report and include:

- oleochemical processes, such as the hydroprocessing of lipid feedstocks obtained from oilseed crops, algae or tallow;
- thermochemical processes, such as the thermochemical conversion of biomass to fluid intermediates (gas or oil) followed by catalytic upgrading and hydroprocessing to hydrocarbon fuels; and
- biochemical processes, such as the biological conversion of biomass (sugars, starches or lignocellulose-derived feedstocks) to longer chain alcohols and hydrocarbons.

A fourth category is also briefly described that includes “hybrid” thermochemical/biochemical technologies such as fermentation of synthesis gas and catalytic reforming of sugars/carbohydrates.

The report concludes by stating:

“Tremendous entrepreneurial activity to develop and commercialize drop-in biofuels from aquatic and terrestrial feedstocks has taken place over the past several years. However, despite these efforts, drop-in biofuels represent only a small percentage (around 2%) of global biofuel markets. (...) Due to the increased processing and resource requirements (e.g., hydrogen and catalysts) needed to make drop-in biofuels as compared to conventional biofuels, large scale production of cost-competitive drop-in biofuels is not expected to occur in the near to midterm. Rather, dedicated policies to promote development and commercialization of these fuels will be needed before they become significant contributors to global biofuels production. Currently, no policies (e.g., tax breaks, subsidies etc.) differentiate new, more fungible and infrastructure ready drop-in type biofuels from less infrastructure compatible oxygenated biofuels. (...) Thus, while tremendous technical progress has been made in developing and improving the various routes to drop-in fuels, supportive policies directed specifically towards the further development of drop-in biofuels are likely to be needed to ensure their future commercial success”

References

- The Royal Society (January 2008). Sustainable biofuels: prospects and challenges, ISBN 978-0-85403-662-2, p. 61.
- Organosolv delignification of willow (2010), AL Macfarlane, MM Farid, JJJ Chen, Lambert Academic Press, ISBN 978-3-8383-9155-7
- Swedish Motor Fuel Technology Co. (1986). Alcohols and alcohol blends as motor fuels, Vol II A & Vol II B. State-of-the-Art" report. Swedish National Board for Technical Development. ISBN 91-7850-156-3.
- Burton, George; Holman, John; Lazonby, John (2000). Salters Advanced Chemistry: Chemical Storylines (2nd ed.). Heinemann. ISBN 0-435-63119-5
- Inslee, Jay; Bracken Hendricks (2007). "6. Homegrown Energy". Apollo's Fire. Island Press, Washington, D.C. pp. 153–155, 160–161. ISBN 978-1-59726-175-3.
- Reuters (2008-09-04). "Brazil SP cane growers to ban burning by 2017". UK Yahoo News. Retrieved 2008-09-14.
- "Biogas Flows Through Germany's Grid Big Time - Renewable Energy News Article". 14 March 2012. Archived from the original on 14 March 2012. Retrieved 17 June 2016.
- Seite nicht gefunden auf Server der Fachagentur Nachwachsende Rohstoffe e.V.: FNR" (PDF). Retrieved 17 June 2016.
- Johnson, Eric (18 Feb 2016). "Integrated enzyme production lowers the cost of cellulosic ethanol". Biofuels, Bioproducts & Biorefining. doi:10.1002/bbb.1634. Retrieved 3 June 2016.
- Spakowicz, Daniel J.; Strobel, Scott A. (2015). "Biosynthesis of hydrocarbons and volatile organic compounds by fungi: bioengineering potential". Applied microbiology and biotechnology. 99 (12): 4943–4951. Retrieved 2016-02-22.
- "Publications - International Resource Panel". Archived from the original on 1 January 2016. Retrieved 30 May 2015.

3

Technologies Related to Biodiesel Production

One of the techniques related to biodiesel production is the mixing of alcohol fuels with biomass. It is done by using MixAlco process. Biomass heating systems helps in generating heat from biomass. The aspects elucidated in this text are of vital importance, and provide a better understanding of biodiesel production.

Bioconversion of Biomass to Mixed Alcohol Fuels

The bioconversion of biomass to mixed alcohol fuels can be accomplished using the MixAlco process. Through bioconversion of biomass to a mixed alcohol fuel, more energy from the biomass will end up as liquid fuels than in converting biomass to ethanol by yeast fermentation.

The process involves a biological/chemical method for converting any biodegradable material (e.g., urban wastes, such as municipal solid waste, biodegradable waste, and sewage sludge, agricultural residues such as corn stover, sugarcane bagasse, cotton gin trash, manure) into useful chemicals, such as carboxylic acids (e.g., acetic, propionic, butyric acid), ketones (e.g., acetone, methyl ethyl ketone, diethyl ketone) and biofuels, such as a mixture of primary alcohols (e.g., ethanol, propanol, *n*-butanol) and/or a mixture of secondary alcohols (e.g., isopropanol, 2-butanol, 3-pentanol). Because of the many products that can be economically produced, this process is a true biorefinery.

Pilot Plant (College Station, Texas)

The process uses a mixed culture of naturally occurring microorganisms found in natural habitats such as the rumen of cattle, termite guts, and marine and terrestrial swamps to anaerobically digest biomass into a mixture of carboxylic acids produced during the acidogenic and acetogenic stages of anaerobic digestion, however with the inhibition of the methanogenic final stage. The more popular methods for production of ethanol and cellulosic ethanol use enzymes that must be isolated first to be added to the biomass and thus convert the starch or cellulose into simple sugars, followed then by yeast fermentation into ethanol. This process does not need the addition of such enzymes as these microorganisms make their own.

As the microoganisms anaerobically digest the biomass and convert it into a mixture of carboxylic acids, the pH must be controlled. This is done by the addition of a buffering agent (e.g., ammonium bicarbonate, calcium carbonate), thus yielding a mixture of carboxylate salts. Methanogenesis, being the natural final stage of anaerobic digestion, is inhibited by the presence of the ammonium ions or by the addition of an inhibitor (e.g., iodoform). The resulting fermentation broth contains the produced carboxylate salts that must be dewatered. This is achieved efficiently by vapor-compression evaporation. Further chemical refining of the dewatered fermentation broth may then take place depending on the final chemical or biofuel product desired.

The condensed distilled water from the vapor-compression evaporation system is recycled back to the fermentation. On the other hand, if raw sewage or other waste water with high BOD in need of treatment is used as the water for the fermentation, the condensed distilled water from the evaporation can be recycled back to the city or to the original source of the high-BOD waste water. Thus, this process can also serve as a water treatment facility, while producing valuable chemicals or biofuels.

Because the system uses a mixed culture of microorganisms, besides not needing any enzyme addition, the fermentation requires no sterility or aseptic conditions, making this front step in the process more economical than in more popular methods for the production of cellulosic ethanol. These savings in the front end of the process, where volumes are large, allows flexibility for further chemical transformations after dewatering, where volumes are small.

Carboxylic Acids

Carboxylic acids can be regenerated from the carboxylate salts using a process known as "acid springing". This process makes use of a high-molecular-weight tertiary amine (e.g., trioctylamine), which is switched with the cation (e.g., ammonium or calcium). The resulting amine carboxylate can then be thermally decomposed into the amine itself, which is recycled, and the corresponding carboxylic acid. In this way, theoretically, no chemicals are consumed or wastes produced during this step.

Ketones

There are two methods for making ketones. The first one consists on thermally converting calcium carboxylate salts into the corresponding ketones. This was a common method for making acetone from calcium acetate during World War I. The other method for making ketones consists on converting the vaporized carboxylic acids on a catalytic bed of zirconium oxide.

Alcohols

Primary Alcohols

The undigested residue from the fermentation may be used in gasification to make hydrogen (H_2). This H_2 can then be used to hydrogenolyze the esters over a catalyst (e.g., copper chromite), which are produced by esterifying either the ammonium carboxylate salts (e.g., ammonium acetate, propionate, butyrate) or the carboxylic acids (e.g., acetic, propionic, butyric acid) with a high-molecular-weight alcohol (e.g., hexanol, heptanol). From the hydrogenolysis, the final products are the high-molecular-weight alcohol, which is recycled back to the esterification, and the corresponding primary alcohols (e.g., ethanol, propanol, butanol).

Secondary Alcohols

The secondary alcohols (e.g., isopropanol, 2-butanol, 3-pentanol) are obtained by hydrogenating over a catalyst (e.g., Raney nickel) the corresponding ketones (e.g., acetone, methyl ethyl ketone, diethyl ketone).

Drop-in Biofuels

The primary or secondary alcohols obtained as described above may undergo conversion to drop-in biofuels, fuels which are compatible with current fossil fuel infrastructure such as biogasoline, green diesel and bio-jet fuel. Such is done by subjecting the alcohols to dehydration followed by oligomerization using zeolite catalysts in a manner similar to the methanex process, which used to produce gasoline from methanol in New Zealand.

Acetic Acid Versus Ethanol

Cellulosic-ethanol manufacturing plants are bound to be net exporters of electricity because a large portion of the lignocellulosic biomass, namely lignin, remains undigested and it must be burned, thus producing electricity for the plant and excess electricity for the grid. As the market grows and this technology becomes more widespread, coupling the liquid fuel and the electricity markets will become more and more difficult.

Acetic acid, unlike ethanol, is biologically produced from simple sugars without the production of carbon dioxide:

$$C_6H_{12}O_6 \rightarrow 2\,CH_3CH_2OH + 2\,CO_2$$

(Biological production of ethanol)

$$C_6H_{12}O_6 \rightarrow 3\,CH_3COOH$$

(Biological production of acetic acid)

Because of this, on a mass basis, the yields will be higher than in ethanol fermentation. If then, the undigested residue (mostly lignin) is used to produce hydrogen by gasification, it is ensured that more energy from the biomass will end up as liquid fuels rather than excess heat/electricity.

$$3\,CH_3COOH + 6\,H_2 \rightarrow 3\,CH_3CH_2OH + 3\,H_2O$$

(Hydrogenation of acetic acid)

$$C_6H_{12}O_6 \text{ (from cellulose)} + 6\,H_2 \text{ (from lignin)} \rightarrow 3\,CH_3CH_2OH + 3\,H_2O$$

(Overall reaction)

A more comprehensive description of the economics of each of the fuels is given on the pages alcohol fuel and ethanol fuel, more information about the economics of various systems can be found on the central page biofuel.

Stage of Development

The system has been in development since 1991, moving from the laboratory scale (10 g/day) to the pilot scale (200 lb/day) in 2001. A small demonstration-scale plant (5 ton/day) has been constructed and is under operation and a 220 ton/day demonstration plant is expected in 2012.

Biomass Heating System

Wood chips in a storage hopper, in the middle an agitator to transport the material with a screw conveyor to the boiler

Biomass heating systems generate heat from biomass.The systems fall under the categories of:

- direct combustion,
- gasification,
- combined heat and power (CHP),
- anaerobic digestion,
- aerobic digestion.

Benefits of Biomass Heating

The use of biomass in heating systems is beneficial because it uses agricultural, forest, urban and industrial residues and waste to produce heat and electricity with less effect on the environment than fossil fuels. This type of energy production has a limited long-term effect on the environment because the carbon in biomass is part of the natural carbon cycle; while the carbon in fossil fuels is not, and permanently adds carbon to the environment when burned for fuel (carbon footprint). Historically, before the use of fossil fuels in significant quantities, biomass in the form of wood fuel provided most of humanity's heating.

Drawbacks of Biomass Heating

On a large scale, the use of biomass takes agricultural land out of food production, reduces the carbon sequestration capacity of forests, and extracts nutrients from the soil. Combustion of biomass creates air pollutants and adds significant quantities of carbon to the atmosphere that may not be returned to the soil for many decades.

Using biomass as a fuel produces air pollution in the form of carbon monoxide, NOx (nitrogen oxides), VOCs (volatile organic compounds), particulates and other pollutants, in some cases at levels above those from traditional fuel sources such as coal or natural gas. Black carbon – a pollutant created by incomplete combustion of fossil fuels, biofuels, and biomass – is possibly the second largest contributor to global warming. In 2009 a Swedish study of the giant brown haze that periodically covers large areas in South Asia determined that it had been principally produced by biomass burning, and to a lesser extent by fossil-fuel burning. Researchers measured a significant concentration of C, which is associated with recent plant life rather than with fossil fuels.

On combustion, the carbon from biomass is released into the atmosphere as carbon dioxide (CO_2). The amount of carbon stored in dry wood is approximately 50% by weight. When from agricultural sources, plant matter used as a fuel can be replaced by planting for new growth. When the biomass is from forests, the time to recapture the carbon stored is generally longer, and the carbon storage capacity of the forest may be reduced overall if destructive forestry techniques are employed.

The biomass-is-carbon-neutral proposal put forward in the early 1990s has been superseded by more recent science that recognizes that mature, intact forests sequester carbon more effectively than cut-over areas. When a tree's carbon is released into the atmosphere in a single pulse, it contributes to climate change much more than woodland timber rotting slowly over decades. Current studies indicate that "even after 50 years the forest has not recovered to its initial carbon storage" and "the optimal strategy is likely to be protection of the standing forest".

Biomass Heating in Our World

The oil price increases since 2003 and consequent price increases for natural gas and coal have increased the value of biomass for heat generation. Forest renderings, agricultural waste, and crops grown specifically for energy production become competitive as the prices of energy dense fossil fuels rise. Efforts to develop this potential may have the effect of regenerating mismanaged croplands and be a cog in the wheel of a decentralized, multi-dimensional renewable energy industry. Efforts to promote and advance these methods became common throughout the European Union through the 2000s. In other areas of the world, inefficient and polluting means to generate heat from biomass coupled with poor forest practices have significantly added to environmental degradation.

Biomass Heating in Europe

European heat market has high level of diversification, which is achieved by offering and using many different solutions in order to satisfy the demands of the final consumer at an acceptable cost. Heat is mainly produced from primary energy supply by means of calorific value conversion, which is used for various kinds of fuels such as coal, fuel oil, natural gas and biomass. Considerable amounts of heat are retrieved from industrial, power and waste incineration processes. This heat is normally produced far from the heat demands. Germany has had the lead until now, its market being the largest one, with revenue share mounting up to 30, 9 per cent in 2000. In spite of this, there has been a continuous drop of this percentage share over the course of the next three years - and this tendency has remained to these days. Due to the fact that Germany has a rather stable market. The next largest markets, with revenue shares in the amount of 22, 3 and 19, 1 per cent, belong to Italy and France respectively. However, Italy has also experienced a strong decline since 2000 while the market situation in France is directly the opposite: its market reached a 20, 2 per cent revenue share by 2006 and became the larger of the two. As for the United Kingdom, it has a rather unchanging and steady market, its growth and rates being linked directly to those of the total European market as a whole

Buffer Tanks

Buffer tanks, sometimes referred to as 'thermal stores' are crucial for the efficient operation of all biomass boilers where the system loading fluctuates rapidly, or the volume of water in the complete hydraulic system is relatively small. Using a suitably sized buffer vessel prevents rapid cycling of the boiler when the loading is below the minimum boiler output. Rapid cycling of the boiler causes a large increase in harmful emissions such as Carbon monoxide, dust, and NOx, greatly reduces boiler efficiency and increases electrical consumption of the unit. In addition, service and maintenance requirements will be increased as parts are stressed by rapid heating and cooling cycles. Although most boilers claim to be able to turn down to 30% of nominal output, in the real world this is often not achievable due to differences in the fuel from the 'ideal' or test fuel. A suitably sized buffer tank should therefore be considered where the loading of the boiler drops below 50% of the nominal output – in other words unless the biomass component is purely base load, the system should include a buffer tank. In any case where the secondary system does not contain sufficient water for safe removal of residual heat from the biomass boiler irrespective of the loading conditions, the system must include a suitably sized buffer tank. The residual heat from a biomass unit varies greatly depending on the boiler design and the thermal mass of the combustion chamber. light weight, fast response boilers require only 10ltr/kW, while industrial wet wood units with very high thermal mass require 40ltr/kW

Types of Biomass Heating Systems

Little biomass heating plant in Austria; the heat power is about 1000 kW

The use of Biomass in heating systems has a use in many different types of buildings, and all have different uses. There are four main types of heating systems that use biomass to heat a boiler. The types are Fully Automated, Semi-Automated, Pellet-Fired, and Combined Heat and Power.

Fully Automated

Fully automated systems operate exactly how they sound. Chipped or ground up waste wood is brought to the site by delivery trucks and dropped into a holding tank. A system of conveyors then transports the wood from the holding tank to the boiler at a certain managed rate. This rate is managed by computer controls and a laser that measures the load of fuel the conveyor is bringing in. The system automatically goes on and off to maintain the pressure and temperature within the boiler. Fully automated systems offer a great deal of ease in their operation because they only require the operator of the system to control the computer, and not the transport of wood.

Semi-automated or "Surge Bin"

Semi-Automated or "Surge Bin" systems are very similar to fully automated systems except they require more manpower to keep operational. They have smaller holding tanks, and a much simpler conveyor systems which will require personnel to maintain the systems operation. The reasoning for the changes from the fully automated system is the efficiency of the system. Wood fire fueled boilers are most efficient when they are running at their highest capacity, and the heat required most days of the year will not be the peak heat requirement for the year. Considering that the system will only need to run at a high capacity a few days of the year, it is made to meet the requirements for the majority of the year to maintain its high efficiency.

Pellet-fired

The third main type of biomass heating systems are pellet-fired systems. Pellets are a processed form of wood, which make them more expensive. Although they are more expensive, they are much more condensed and uniform, and therefore are more efficient. Further, it is relatively easy to automatically feed pellets to boilers. In these systems, the pellets are stored in a grain-type storage silo, and gravity is used to move them to the boiler. The storage requirements are much smaller for pellet-fired systems because of their condensed nature, which also helps cut down costs. these systems are used for a wide variety of facilities, but they are most efficient and cost effective for places where space for storage and conveyor systems is limited, and where the pellets are made fairly close to the facility.

Agricultural Pellet Systems

Sub category of pellet systems are boilers or burners capable of burning pellet with higher ash rate (paper pellets, hay pellets, straw pellets). One of this kind is PETROJET pellet burner with rotating cylindrical burning chamber. In terms of efficiencies advanced pellet boilers can exceed other forms of biomass because of the more stable fuel charataristics. Advanced pellet boilers can even work in condensing mode and cool down combustion gases to 30-40 °C, instead of 120 °C before sent into the flue.

Combined Heat and Power

Combined heat and power systems are very useful systems in which wood waste, such as wood chips, is used to generate power, and heat is created as a byproduct of the power generation system. They have a very high cost because of the high pressure operation. Because of this high pressure operation, the need for a highly trained operator is mandatory, and will raise the cost of operation. Another drawback is that while they produce electricity they will produce heat, and if producing heat is not desirable for certain parts of the year, the addition of a cooling tower is necessary, and will also raise the cost.

There are certain situations where CHP is a good option. Wood product manufacturers would use a combined heat and power system because they have a large supply of waste wood, and a need for both heat and power. Other places where these systems would be optimal are hospitals and prisons, which need energy, and heat for hot water. These systems are sized so that they will produce enough heat to match the average heat load so that no additional heat is needed, and a cooling tower is not needed.

Gasification

Part of a series about

Sustainable energy

Gasification is a process that converts organic or fossil fuel based carbonaceous materials into carbon monoxide, hydrogen and carbon dioxide. This is achieved by reacting the material at high temperatures (>700 °C), without combustion, with a controlled amount of oxygen and/or steam. The resulting gas mixture is called syngas (from synthesis gas or synthetic gas) or producer gas and is itself a fuel. The power derived from gasification and combustion of the resultant gas is considered to be a source of renewable energy if the gasified compounds were obtained from biomass.

The advantage of gasification is that using the syngas is potentially more efficient than direct combustion of the original fuel because it can be combusted at higher temperatures or even in fuel cells, so that the thermodynamic upper limit to the efficiency defined by Carnot's rule is higher or (in case of fuel cells) not applicable. Syngas may be burned directly in gas engines, used to produce methanol and hydrogen, or converted via the Fischer–Tropsch process into synthetic fuel. Gasification can also begin with material which would otherwise have been disposed of such as biodegradable waste. In addition, the high-temperature process refines out corrosive ash elements such as chloride and potassium, allowing clean gas production from otherwise problematic fuels. Gasification of fossil fuels is currently widely used on industrial scales to generate electricity.

History

Adler Diplomat 3 with gas generator (1941)

The process of producing energy using the gasification method has been in use for more than 180 years. In the early time coal and peat were used to power these plants. Initially developed to produce town gas for lighting and cooking in the 1800s, this was replaced by electricity and natural gas, it was also used in blast furnaces but the bigger role was played in the production of synthetic chemicals where it has been in use since the 1920s.

During both world wars, especially the World War II, the need for fuel produced by gasification reemerged due to the shortage of petroleum. Wood gas generators, called Gasogene or Gazogène, were used to power motor vehicles in Europe. By 1945 there were trucks, buses and agricultural machines that were powered by gasification. It is estimated that there were close to 9,000,000 vehicles running on producer gas all over the world.

Chemical Reactions

In a gasifier, the carbonaceous material undergoes several different processes:

1. The dehydration or drying process occurs at around 100 °C. Typically the resulting steam is mixed into the gas flow and may be involved with subsequent

chemical reactions, notably the water-gas reaction if the temperature is sufficiently high.

2. The *pyrolysis* (or devolatilization) process occurs at around 200-300 °C. Volatiles are released and char is produced, resulting in up to 70% weight loss for coal. The process is dependent on the properties of the carbonaceous material and determines the structure and composition of the char, which will then undergo gasification reactions.

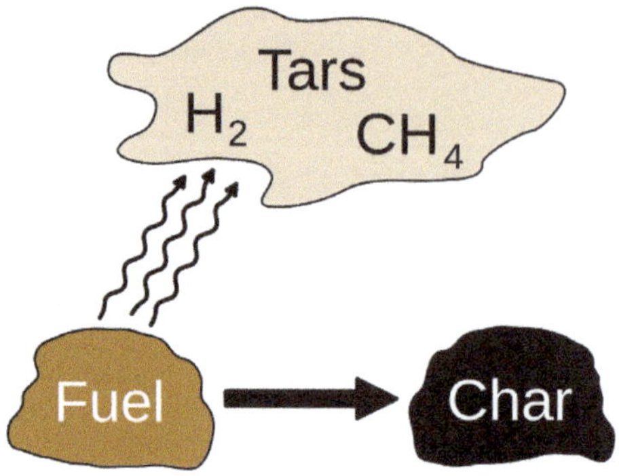

Pyrolysis of carbonaceous fuels

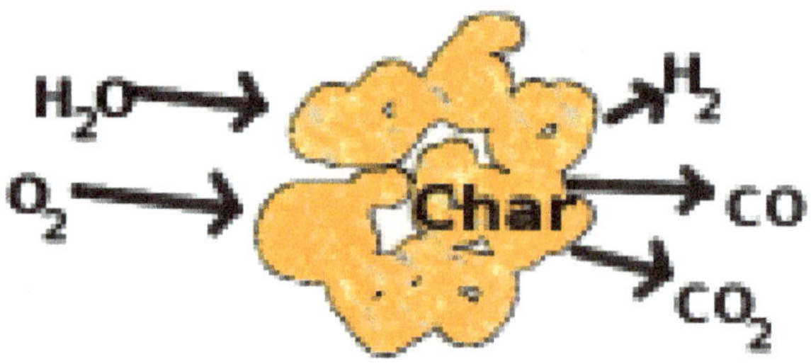

Gasification of char

3. The *combustion* process occurs as the volatile products and some of the char react with oxygen to primarily form carbon dioxide and small amounts of carbon monoxide, which provides heat for the subsequent gasification reactions. Letting **C** represent a carbon-containing organic compound, the basic reaction here is $C + O2 \rightarrow CO2$

4. The *gasification* process occurs as the char reacts with steam to produce carbon monoxide and hydrogen, via the reaction $C + H2O \rightarrow H2 + CO$

5. In addition, the reversible gas phase water-gas shift reaction reaches equilibrium very fast at the temperatures in a gasifier. This balances the concentrations of carbon monoxide, steam, carbon dioxide and hydrogen. $C O + H 2 O \leftrightarrow C O 2 + H 2$

6. In essence, a limited amount of oxygen or air is introduced into the reactor to allow some of the organic material to be "burned" to produce carbon dioxide and energy, which drives a second reaction that converts further organic materi-

al to hydrogen and additional carbon dioxide. Further reactions occur when the formed carbon monoxide and residual water from the organic material react to form methane and excess carbon dioxide ($4CO + 2H_2O \rightarrow CH_4 + 3CO_2$). This third reaction occurs more abundantly in reactors that increase the residence time of the reactive gases and organic materials, as well as heat and pressure. Catalysts are used in more sophisticated reactors to improve reaction rates, thus moving the system closer to the reaction equilibrium for a fixed residence time.

Gasification Processes

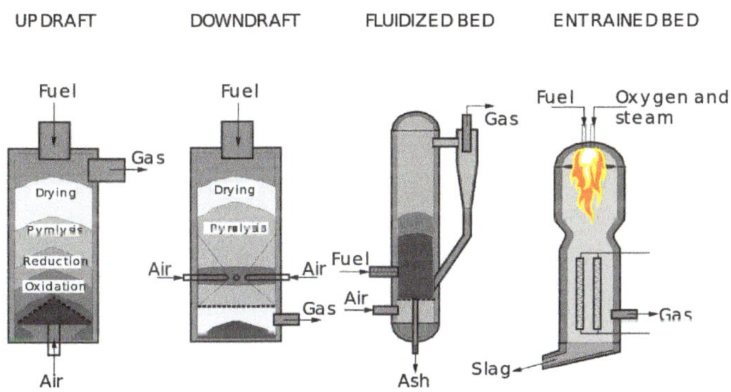

Main gasifier types

Several types of gasifiers are currently available for commercial use: counter-current fixed bed, co-current fixed bed, fluidized bed, entrained flow, plasma, and free radical.

Counter-current Fixed Bed ("Up Draft") Gasifier

A fixed bed of carbonaceous fuel (e.g. coal or biomass) through which the "gasification agent" (steam, oxygen and/or air) flows in counter-current configuration. The ash is either removed in the dry condition or as a slag. The slagging gasifiers have a lower ratio of steam to carbon, achieving temperatures higher than the ash fusion temperature. The nature of the gasifier means that the fuel must have high mechanical strength and must ideally be non-caking so that it will form a permeable bed, although recent developments have reduced these restrictions to some extent. The throughput for this type of gasifier is relatively low. Thermal efficiency is high as the temperatures in the gas exit are relatively low. However, this means that tar and methane production is significant at typical operation temperatures, so product gas must be extensively cleaned before use. The tar can be recycled to the reactor.

In the gasification of fine, undensified biomass such as rice hulls, it is necessary to blow air into the reactor by means of a fan. This creates very high gasification temperature, as high as 1000 C. Above the gasification zone, a bed of fine and hot char is formed, and as the gas is blow forced through this bed, most complex hydrocarbons are broken down into simple components of hydrogen and carbon monoxide.

Co-current Fixed Bed ("Down Draft") Gasifier

Similar to the counter-current type, but the gasification agent gas flows in co-current configuration with the fuel (downwards, hence the name "down draft gasifier"). Heat needs to be added to the upper part of the bed, either by combusting small amounts of the fuel or from external heat sources. The produced gas leaves the gasifier at a high temperature, and most of this heat is often transferred to the gasification agent added in the top of the bed, resulting in an energy efficiency on level with the counter-current type. Since all tars must pass through a hot bed of char in this configuration, tar levels are much lower than the counter-current type.

Fluidized Bed Reactor

The fuel is fluidized in oxygen and steam or air. The ash is removed dry or as heavy agglomerates that defluidize. The temperatures are relatively low in dry ash gasifiers, so the fuel must be highly reactive; low-grade coals are particularly suitable. The agglomerating gasifiers have slightly higher temperatures, and are suitable for higher rank coals. Fuel throughput is higher than for the fixed bed, but not as high as for the entrained flow gasifier. The conversion efficiency can be rather low due to elutriation of carbonaceous material. Recycle or subsequent combustion of solids can be used to increase conversion. Fluidized bed gasifiers are most useful for fuels that form highly corrosive ash that would damage the walls of slagging gasifiers. Biomass fuels generally contain high levels of corrosive ash.

Entrained Flow Gasifier

A dry pulverized solid, an atomized liquid fuel or a fuel slurry is gasified with oxygen (much less frequent: air) in co-current flow. The gasification reactions take place in a dense cloud of very fine particles. Most coals are suitable for this type of gasifier because of the high operating temperatures and because the coal particles are well separated from one another.

The high temperatures and pressures also mean that a higher throughput can be achieved, however thermal efficiency is somewhat lower as the gas must be cooled before it can be cleaned with existing technology. The high temperatures also mean that tar and methane are not present in the product gas; however the oxygen requirement is higher than for the other types of gasifiers. All entrained flow gasifiers remove the major part of the ash as a slag as the operating temperature is well above the ash fusion temperature.

A smaller fraction of the ash is produced either as a very fine dry fly ash or as a black colored fly ash slurry. Some fuels, in particular certain types of biomasses, can form slag that is corrosive for ceramic inner walls that serve to protect the gasifier outer wall. However some entrained flow type of gasifiers do not possess a ceramic inner wall but have an inner water or steam cooled wall covered with partially solidified slag. These types of gasifiers do not suffer from corrosive slags.

Some fuels have ashes with very high ash fusion temperatures. In this case mostly limestone is mixed with the fuel prior to gasification. Addition of a little limestone will usually suffice for the lowering the fusion temperatures. The fuel particles must be much smaller than for other types of gasifiers. This means the fuel must be pulverized, which requires somewhat more energy than for the other types of gasifiers. By far the most energy consumption related to entrained flow gasification is not the milling of the fuel but the production of oxygen used for the gasification.

Plasma Gasifier

In a plasma gasifier a high-voltage current is fed to a torch, creating a high-temperature arc. The inorganic residue is retrieved as a glass-like substance.

Feedstock

There are a large number of different feedstock types for use in a gasifier, each with different characteristics, including size, shape, bulk density, moisture content, energy content, chemical composition, ash fusion characteristics, and homogeneity of all these properties. Coal and petroleum coke are used as primary feedstocks for many large gasification plants worldwide. Additionally, a variety of biomass and waste-derived feedstocks can be gasified, with wood pellets and chips, waste wood, plastics and aluminium, Municipal Solid Waste (MSW), Refuse-derived fuel (RDF), agricultural and industrial wastes, sewage sludge, switch grass, discarded seed corn, corn stover and other crop residues all being used.

Chemrec has developed a process for gasification of black liquor.

Waste Disposal

Waste gasification has several advantages over incineration:

- The necessary extensive flue gas cleaning may be performed on the syngas instead of the much larger volume of flue gas after combustion.
- Electric power may be generated in engines and gas turbines, which are much cheaper and more efficient than the steam cycle used in incineration. Even fuel cells may potentially be used, but these have rather severe requirements regarding the purity of the gas.
- Chemical processing (Gas to liquids) of the syngas may produce other synthetic fuels instead of electricity.
- Some gasification processes treat ash containing heavy metals at very high temperatures so that it is released in a glassy and chemically stable form.

A major challenge for waste gasification technologies is to reach an acceptable (positive) gross electric efficiency. The high efficiency of converting syngas to electric power is counteracted by significant power consumption in the waste preprocessing, the consumption of large amounts of pure oxygen (which is often used as gasification agent), and gas cleaning. Another challenge becoming apparent when implementing the processes in real life is to obtain long service intervals in the plants, so that it is not necessary to close down the plant every few months for cleaning the reactor.

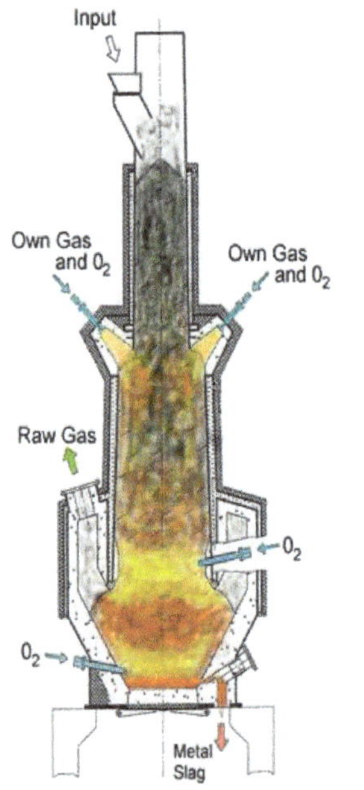

HTCW reactor, one of several proposed waste gasification processes. According to the sales and sales management consultants *KBI Group* a pilot plant in Arnstadt implementing this process has completed initial tests.

Environmental advocates have called gasification "incineration in disguise" and argue that the technology is still dangerous to air quality and public health. "Since 2003 numerous proposals for waste treatment facilities hoping to use... gasification technologies failed to receive final approval to operate when the claims of project proponents did not withstand public and governmental scrutiny of key claims," according to the Global Alliance for Incinerator Alternatives. One facility which operated from 2009-2011 in Ottawa had 29 "emissions incidents" and 13 "spills" over those three years. It was also only able to operate roughly 25% of the time.

Several waste gasification processes have been proposed, but few have yet been built and tested, and only a handful have been implemented as plants processing real waste, and most of the time in combination with fossil fuels.

One plant (in Chiba, Japan using the Thermoselect process) has been processing industrial waste since year 2000, but has not yet documented positive net energy production from the process.

In the United States, gasification of waste is expanding across the country. Ze-gen is operating a waste gasification demonstration facility in New Bedford, Massachusetts. The facility was designed to demonstrate gasification of specific non-MSW waste streams using *liquid metal gasification*. This facility came after widespread public opposition shelved plans for a similar plant in Attleboro, Massachusetts. In addition, construction of a biomass gasification plant was approved in DeKalb County, Georgia on June 14, 2011.

Also in the USA, plasma is being used to gasify municipal solid waste, hazardous waste and biomedical waste at the Hurlburt Field Florida Special Operations Command Air Force base. PyroGenesis Canada Inc. is the technology provider. InnerPoint Energy in Cedar Hill, MO has a gasification system going.

Current Applications

Syngas can be used for heat production and for generation of mechanical and electrical power. Like other gaseous fuels, producer gas gives greater control over power levels when compared to solid fuels, leading to more efficient and cleaner operation.

Syngas can also be used for further processing to liquid fuels or chemicals.

Heat

Gasifiers offer a flexible option for thermal applications, as they can be retrofitted into existing gas fueled devices such as ovens, furnaces, boilers, etc., where syngas may replace fossil fuels. Heating values of syngas are generally around 4-10 MJ/m^3.

Electricity

Currently Industrial-scale gasification is primarily used to produce electricity from fossil fuels such as coal, where the syngas is burned in a gas turbine. Gasification is also used industrially in the production of electricity, ammonia and liquid fuels (oil) using Integrated Gasification Combined Cycles (IGCC), with the possibility of producing methane and hydrogen for fuel cells. IGCC is also a more efficient method of CO_2 capture as compared to conventional technologies. IGCC demonstration plants have been operating since the early 1970s and some of the plants constructed in the 1990s are now entering commercial service.

Combined Heat and Power

In small business and building applications, where the wood source is sustainable, 250-1000 kWe and new zero carbon biomass gasification plants have been installed in Europe that produce tar free syngas from wood and burn it in reciprocating engines connected to a generator with heat recovery. This type of plant is often referred to as a wood biomass CHP unit but is a plant with seven different processes: biomass processing, fuel delivery, gasification, gas cleaning, waste disposal, electricity generation and heat recovery.

Transport Fuel

Diesel engines can be operated on dual fuel mode using producer gas. Diesel substitution of over 80% at high loads and 70-80% under normal load variations can easily be achieved. Spark ignition engines and solid oxide fuel cells can operate on 100% gasification gas. Mechanical energy from the engines may be used for e.g. driving water pumps for irrigation or for coupling with an alternator for electrical power generation.

While small scale gasifiers have existed for well over 100 years, there have been few sources to obtain a ready to use machine. Small scale devices are typically DIY projects. However, currently in the United States, several companies offer gasifiers to operate small engines.

Renewable Energy and Fuels

In principle, gasification can proceed from just about any organic material, including biomass and plastic waste. The resulting syngas can be combusted. Alternatively, if the syngas is clean enough, it may be used for power production in gas engines, gas turbines or even fuel cells, or converted efficiently to dimethyl ether (DME) by methanol dehydration, methane via the Sabatier reaction, or diesel-like synthetic fuel via the Fischer–Tropsch process. In many gasification processes most of the inorganic components of the input material, such as metals and minerals, are retained in the ash. In some gasification processes (slagging gasification) this ash has the form of a glassy solid with low leaching properties, but the net power production in slagging gasification is low (sometimes negative) and costs are higher.

Gasification plant Güssing, Austria (2006)

Regardless of the final fuel form, gasification itself and subsequent processing neither directly emits nor traps greenhouse gases such as carbon dioxide. Power consumption in the gasification and syngas conversion processes may be significant though, and may indirectly cause CO_2 emissions; in slagging and plasma gasification, the electricity consumption may even exceed any power production from the syngas.

Combustion of syngas or derived fuels emits exactly the same amount of carbon dioxide as would have been emitted from direct combustion of the initial fuel. Biomass gasification and combustion could play a significant role in a renewable energy economy, because biomass production removes the same amount of CO_2 from the atmosphere as is emitted from gasification and combustion. While other biofuel technologies such as biogas and biodiesel are carbon neutral, gasification in principle may run on a wider variety of input materials and can be used to produce a wider variety of output fuels.

There are at present a few industrial scale biomass gasification plants. Since 2008 in Svenljunga, Sweden, a biomass gasification plant generates up to 14 MW_{th}, supplying industries and citizens of Svenljunga with process steam and district heating, respectively. The gasifier uses biomass fuels such as CCA or creosote impregnated waste wood and other kinds of recycled wood to produces syngas that is combusted on site. In 2011 a similar gasifier, using the same kinds of fuels, is being installed at Munkfors Energy's CHP plant. The CHP plant will generate 2 MW_e (electricity) and 8 MW_{th} (district heating).

Examples of demonstration projects include:

- The 32 MW dual fluidized bed gasification of the GoBiGas project in Gothenburg, Sweden, producing around 20 MW of substitute natural gas from forest residues and feeding into the natural gas grid since December 2014.
- Those of the Renewable Energy Network Austria, including a plant using dual fluidized bed gasification that has supplied the town of Güssing with 2 MW of electricity, produced utilising GE Jenbacher reciprocating gas engines and 4 MW of heat, generated from wood chips, since 2003.
- Chemrec's pilot plant in Piteå that has produced 3 MW of clean syngas since 2006, generated from entrained flow gasification of black liquor.
- The US Air Force Transportable Plasma Waste to Energy System (TPWES) facility at Hurlburt Field, Florida.

Biorefinery

A biorefinery is a facility that integrates biomass conversion processes and equipment to produce fuels, power, heat, and value-added chemicals from biomass. The biorefinery concept is analogous to today's petroleum refinery, which produce multiple fuels and products from petroleum.

The International Energy Agency Bioenergy Task 42 on Biorefineries has defined biorefining as the sustainable processing of biomass into a spectrum of bio-based products (food, feed, chemicals, materials) and bioenergy (biofuels, power and/or heat).

By producing multiple products, a biorefinery takes advantage of the various components in biomass and their intermediates therefore maximizing the value derived from the biomass feedstock. A biorefinery could, for example, produce one or several low-volume, but high-value, chemical or nutraceutical products and a low-value, but high-volume liquid transportation fuel such as biodiesel or bioethanol. At the same time generating electricity and process heat, through combined heat and power (CHP) technology, for its own use and perhaps enough for sale of electricity to the local utility. The high-value products increase profitability, the high-volume fuel helps meet energy needs, and the power production helps to lower energy costs and reduce greenhouse gas emissions from traditional power plant facilities. Although some facilities exist that can be called bio-refineries, the bio-refinery has yet to be fully realized. Future biorefineries may play a major role in producing chemicals and materials that are traditionally produced from petroleum.

Examples

The fully operational Blue Marble Energy company has multiple biorefineries located in Odessa, WA and Missoula, MT.

Canada's first Integrated Biorefinery, developed on anaerobic digestion technology by Himark BioGas is located in Hairy Hill, Alberta. The Biorefinery utilizes Source Separated Organics from the metro Edmonton region, Open Pen Feedlot Manure, and Food Processing Waste.

Several potential biorefinery examples have been proposed, starting from feedstocks such as tobacco, flax straw and the residues from the production of bioethanol.

Chemrec's technology for black liquor gasification and production of second-generation biofuels such as biomethanol or BioDME is integrated with a host pulp mill and utilizes a major sulfate or sulfite process waste product as feedstock.

Thermal Depolymerization

Thermal depolymerization (TDP) is a depolymerization process using hydrous pyrolysis for the reduction of complex organic materials (usually waste products of various sorts, often biomass and plastic) into light crude oil. It mimics the natural geological processes thought to be involved in the production of fossil fuels. Under pressure and heat, long chain polymers of hydrogen, oxygen, and carbon decompose into short-chain petroleum hydrocarbons with a maximum length of around 18 carbons.

Similar Processes

Thermal depolymerisation is similar to other processes which use superheated water as a major step to produce fuels, such as direct Hydrothermal Liquefaction. These are distinct from processes using dry materials to depolymerize, such as pyrolysis. The term Thermochemical Conversion (TCC) has also been used for conversion of biomass to oils, using superheated water, although it is more usually applied to fuel production via pyrolysis. Other commercial scale processes include the "SlurryCarb" process operated by EnerTech, which uses similar technology to decarboxylate wet solid biowaste, which can then be physically dewatered and used as a solid fuel called E-Fuel. The plant was designed to Rialto to process 683 tons of waste per day. However, it failed to perform to design standards and was closed down. The Rialto facility defaulted on its bond payments and is in the process of being liquidated. The Hydro Thermal Upgrading (HTU) process uses superheated water to produce oil from domestic waste. A demonstration plant is due to start up in The Netherlands said to be capable of processing 64 tons of biomass (dry basis) per day into oil. Thermal depolymerisation differs in that it contains a hydrous process followed by an anhydrous cracking / distillation process.

History

Thermal depolymerization is similar to the geological processes that produced the fossil fuels used today, except that the technological process occurs in a timeframe measured in hours. Until recently, the human-designed processes were not efficient enough to serve as a practical source of fuel—more energy was required than was produced.

The first industrial process to obtain gas, diesel fuels and other petroleum products through pyrolysis of coal, tar or biomass was designed and patented in the late 1920s by Fischer-Tropsch. In U. S. patent 2,177,557, issued in 1939, Bergstrom and Cederquist discuss a method for obtaining oil from wood in which the wood is heated under pressure in water with a significant amount of calcium hydroxide added to the mixture. In the early 1970s Herbert R. Appell and coworkers worked with hydrous pyrolysis methods, as exemplified by U. S. patent 3,733,255 (issued in 1973), which discusses the production of oil from sewer sludge and municipal refuse by heating the material in water, under pressure, and in the presence of carbon monoxide.

An approach that exceeded break-even was developed by Illinois microbiologist Paul Baskis in the 1980s and refined over the next 15 years. The technology was finally developed for commercial use in 1996 by Changing World Technologies (CWT). Brian S. Appel (CEO of CWT) took the technology in 2001 and expanded and changed it into what is now referred to as TCP (Thermal Conversion Process), and has applied for and obtained several patents. A Thermal Depolymerization demonstration plant was completed in 1999 in Philadelphia by Thermal Depolymerization, LLC, and the first full-scale commercial plant was constructed in Carthage, Missouri, about 100 yards (91 m) from ConAgra Foods' massive Butterball turkey plant, where it is expected to process about 200 tons of turkey waste into 500 barrels (79 m^3) of oil per day.

Theory and Process

In the method used by CWT, the water improves the heating process and contributes hydrogen to the reactions.

In the Changing World Technologies (CWT) process, the feedstock material is first ground into small chunks, and mixed with water if it is especially dry. It is then fed into a pressure vessel reaction chamber where it is heated at constant volume to around 250 °C. Similar to a pressure cooker (except at much higher pressure), steam naturally raises the pressure to 600 psi (4 MPa) (near the point of saturated water). These conditions are held for approximately 15 minutes to fully heat the mixture, after which the pressure is rapidly released to boil off most of the water. The result is a mix of crude hydrocarbons and solid minerals. The minerals are removed, and the hydrocarbons are sent to a second-stage reactor where they are heated to 500 °C, further breaking down the longer hydrocarbon chains. The hydrocarbons are then sorted by fractional distillation, in a process similar to conventional oil refining.

The CWT company claims that 15 to 20% of feedstock energy is used to provide energy for the plant. The remaining energy is available in the converted product. Working with turkey offal as the feedstock, the process proved to have yield efficiencies of approximately 85%; in other words, the energy contained in the end products of the process is 85% of the energy contained in the inputs to the process (most notably the energy content of the feedstock, but also including electricity for pumps and natural gas or woodgas for heating). If one considers the energy content of the feedstock to be free (i.e., waste material from some other process), then 85 units of energy are made available for every 15 units of energy consumed in process heat and electricity. This means the "Energy Returned on Energy Invested" (EROEI) is (6.67), which is comparable to other energy harvesting processes. Higher efficiencies may be possible with drier and more carbon-rich feedstocks, such as waste plastic.

By comparison, the current processes used to produce ethanol and biodiesel from agricultural sources have EROEI in the 4.2 range, when the energy used to produce the feedstocks is accounted for (in this case, usually sugar cane, corn, soybeans and the like). These EROEI values are not directly comparable, because these EROEI calculations include the energy cost to produce the feedstock, whereas the above EROEI calculation for thermal depolymerization process (TDP) does not.

The process breaks down almost all materials that are fed into it. TDP even efficiently breaks down many types of hazardous materials, such as poisons and difficult-to-destroy biological agents such as prions.

Feedstocks and Outputs with Thermal Depolymerization

Average TDP Feedstock Outputs				
Feedstock	**Oils**	**Gas-es**	**Solids (mostly carbon based)**	**Water (Steam)**
Plastic bottles	70%	16%	6%	8%
Medical waste	65%	10%	5%	20%
Tires	44%	10%	42%	4%
Turkey offal	39%	6%	5%	50%
Sewage sludge	26%	9%	8%	57%
Paper (cellulose)	8%	48%	24%	20%

(Note: Paper/cellulose contains at least 1% minerals, which was probably grouped under carbon solids.)

Carthage Plant Products

As reported on 04/02/2006 by Discover Magazine, a Carthage, Missouri plant was producing 500 barrels per day (79 m^3/d) of oil made from 270 tons of turkey entrails and 20 tons of hog lard. This represents an oil yield of 22.3 percent. The Carthage plant produces API 40+, a high value crude oil. It contains light and heavy naphthas, a kerosene, and a gas oil fraction, with essentially no heavy fuel oils, tars, asphaltenes or waxes. It can be further refined to produce No. 2 and No. 4 fuel oils.

TDP-40 Oil Classification by D-5443 PONA method	
Output Material	**% by Weight**
Paraffins	22%
Olefins	14%
Naphthenes	3%
Aromatics	6%
C14/C14+	55%
	100%

The fixed carbon solids produced by the TDP process have multiple uses as a filter, a fuel source and a fertilizer. It can be used as activated carbon in wastewater treatment, as a fertilizer, or as a fuel similar to coal.

Advantages

The process can break down organic poisons, due to breaking chemical bonds and destroying the molecular shape needed for the poison's activity. It is likely to be highly effective at killing pathogens, including prions. It can also safely remove heavy metals from the samples by converting them from their ionized or organometallic forms to their stable oxides which can be safely separated from the other products.

Along with similar processes, it is a method of recycling the energy content of organic materials without first removing the water. It can produce liquid fuel, which separates from the water physically without need for drying. Other methods to recover energy often require pre-drying (e.g. burning, pyrolysis) or produce gaseous products (e.g. anaerobic digestion).

Potential Sources of Waste Inputs

The United States Environmental Protection Agency estimates that in 2006 there were 251 million tons of municipal solid waste, or 4.6 pounds generated per day per person in the USA. Much of this mass is considered unsuitable for oil conversion.

Limitations

The process only breaks long molecular chains into shorter ones, so small molecules such as carbon dioxide or methane cannot be converted to oil through this process. However, the methane in the feedstock is recovered and burned to heat the water that is an essential part of the process. In addition, the gas can be burned in a combined heat and power plant, consisting of a gas turbine which drives a generator to create electricity, and a heat exchanger to heat the process input water from the exhaust gas. The electricity can be sold to the power grid, for example under a feed-in tariff scheme. This also increases the overall efficiency of the process (already said to be over 85% of feedstock energy content).

Another option is to sell the methane product as biogas. For example, biogas can be compressed, much like natural gas, and used to power motor vehicles.

Many agricultural and animal wastes could be processed, but many of these are already used as fertilizer, animal feed, and, in some cases, as feedstocks for paper mills or as boiler fuel. Energy crops constitute another potentially large feedstock for thermal depolymerization.

Current Status

Reports in 2004 claimed that the Carthage facility was selling products at 10% below the price of equivalent oil, but its production costs were low enough that it produced a profit. At the time it was paying for turkey waste.

The plant then consumed 270 tons of turkey offal (the full output of the turkey processing plant) and 20 tons of egg production waste daily. In February 2005, the Carthage plant was producing about 400 barrels per day (64 m^3/d) of crude oil.

In April 2005 the plant was reported to be running at a loss. Further 2005 reports summarized some economic setbacks which the Carthage plant encountered since its planning stages. It was thought that concern over mad cow disease would prevent the use of turkey waste and other animal products as cattle feed, and thus this waste would be free. As it turned out, turkey waste may still be used as feed in the United States, so that the facility must purchase that feed stock at a cost of $30 to $40 per ton, adding $15 to $20 per barrel to the cost of the oil. Final cost, as of January 2005, was $80/barrel ($1.90/gal).

The above cost of production also excludes the operating cost of the thermal oxidizer and scrubber added in May 2005 in response to odor complaints.

A biofuel tax credit of roughly $1 per US gallon (26 ¢/L) on production costs was not available because the oil produced did not meet the definition of "biodiesel" according to the relevant American tax legislation. The Energy Policy Act of 2005 specifically added thermal depolymerization to a $1 renewable diesel credit, which became effective at the end of 2005, allowing a profit of $4/barrel of output oil.

Company Expansion

The company has explored expansion in California, Pennsylvania, and Virginia, and is presently examining projects in Europe, where animal products cannot be used as cattle feed. TDP is also being considered as an alternative means for sewage treatment in the United States.

Smell Complaints

The pilot plant in Carthage was temporarily shut down due to smell complaints. It was soon restarted when it was discovered that few of the odors were generated by the plant. Furthermore, the plant agreed to install an enhanced thermal oxidizer and to upgrade its air scrubber system under a court order. Since the plant is located only four blocks from the tourist-attracting town center, this has strained relations with the mayor and citizens of Carthage.

According to a company spokeswoman, the plant has received complaints even on days when it is not operating. She also contended that the odors may not have been produced by their facility, which is located near several other agricultural processing plants.

On December 29, 2005, the plant was ordered by the state governor to shut down once again over allegations of foul odors as reported by MSNBC.

As of March 7, 2006, the plant has begun limited test runs to validate it has resolved the odor issue.

As of August 24, 2006, the last lawsuit connected with the odor issue has been dismissed and the problem is acknowledged as fixed. In late November, however, another complaint was filed over bad smells. This complaint was closed on January 11 of 2007 with no fines assessed.

Status as of February 2009

A May 2003 article in Discover magazine stated, "Appel has lined up federal grant money to help build demonstration plants to process chicken offal and manure in Alabama and crop residuals and grease in Nevada. Also in the works are plants to process turkey waste and manure in Colorado and pork and cheese waste in Italy. He says the first generation of depolymerization centers will be up and running in 2005. By then it should be clear whether the technology is as miraculous as its backers claim."

However, as of August 2008, the only operational plant listed at the company's website is the initial one in Carthage, Missouri.

Changing World Technology applied for an IPO on August 12; 2008, hoping to raise $100 million.

The unusual Dutch Auction type IPO failed possibly because CWT has lost nearly $20 million with very little revenue.

CWT, the parent company of Renewable Energy Solutions, filed for Chapter 11 bankruptcy. No details on plans for the Carthage plant have been released.

In April 2013, CWT was acquired by a Canadian firm, Ridgeline Energy Services, based in Calgary.

Similar Technologies

- Plasma Converters use powerful electric arcs to reduce and extract energy from waste.
- Wet oxidation
- Hydrocracking

References

- "Göteborg Energi's biogas plant GoBiGas is now fully operational - GoBiGas". gobigas.goteborgenergi.se. Retrieved 2015-11-09.
- Dr W J Smith, Tamutech Consultancy. Mapping the Development of UK Biorefinery Complexes, NNFCC, 2007-06-20. Retrieved on 2011-02-16.
- National Non-Food Crops Centre. The Biorefinery Opportunity - A North East England view, NNFCC 08-001, NNFCC, 2008-06-28. Retrieved on 2011-02-16.
- "AFSOC makes 'green' history while investing in future". US Air Force Special Operations Command. Retrieved 2011-04-28.
- Kamka, Frank; Jochmann, Andreas (June 2005). Development Status of BGL-Gasification (PDF). International Freiberg Conference on IGCC & XtL Technologies. speaker Lutz Picard. Retrieved 2011-03-19.

- National Non-Food Crops Centre. "Review of Technologies for Gasification of Biomass and Wastes, NNFCC 09-008", Retrieved on 2011-06-24
- "Treetrouble: Testimonies on the Negative Impact of Large-scale Tree Plantations prepared for the sixth Conference of the Parties of the Framework Convention on Climate Change". Friends of the Earth International. Retrieved 11 August 2010.
- Prasad, Ram. "SUSTAINABLE FOREST MANAGEMENT FOR DRY FORESTS OF SOUTH ASIA". Food and Agriculture Organization of the United Nations. Retrieved 11 August 2010.
- Mary S. Booth. "Biomass Briefing, October 2009" (PDF). massenvironmentalenergy.org. Massachusetts Environmental Energy Alliance. Retrieved 12 December 2010.

4

Methods and Components of Biodiesel Production

The production of biodiesel deals with a number of methods and components. Some of these methods are titration, tallow, fatty acid methyl ester and husk. Titration is a common laboratory method which is used to regulate unknown concentration of identified components. The chapter strategically encompasses and incorporates the major components and methods of biodiesel production, providing a complete understanding.

Titration

Titration, also known as titrimetry, is a common laboratory method of quantitative chemical analysis that is used to determine the unknown concentration of an identified analyte. Since volume measurements play a key role in titration, it is also known as volumetric analysis. A reagent, called the *titrant* or *titrator* is prepared as a standard solution. A known concentration and volume of titrant reacts with a solution of *analyte* or *titrand* to determine concentration. The volume of titrant reacted is called *titration volume*.

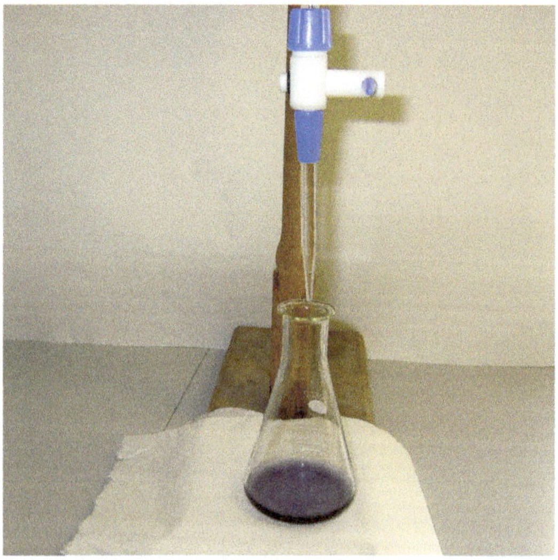

A Winkler titration to determine the concentration of dissolved oxygen in a water sample. The dissolved oxygen has been converted to an equivalent amount of iodine, which is being titrated with thiosulfate using a starch indicator. The blue color in the flask will disappear when all the iodine has been converted to iodide.

History and Etymology

The word "titration" comes from the Latin word *titulus*, meaning inscription or title. The French word *titre*, also from this origin, means rank.

Volumetric analysis originated in late 18th-century France. François-Antoine-Henri Descroizilles (fr) developed the first burette (which was similar to a graduated cylinder) in 1791. Joseph Louis Gay-Lussac developed an improved version of the burette that included a side arm, and coined the terms "pipette" and "burette" in an 1824 paper on the standardization of indigo solutions. A major breakthrough in the methodology and popularization of volumetric analysis was due to Karl Friedrich Mohr, who redesigned the burette by placing a clamp and a tip at the bottom, and wrote the first textbook on the topic, *Lehrbuch der chemisch-analytischen Titrirmethode* (*Textbook of analytical-chemical titration methods*), published in 1855.

Procedure

Analysis of soil samples by titration

A typical titration begins with a beaker or Erlenmeyer flask containing a very precise volume of the analyte and a small amount of indicator (such as phenolphthalein) placed underneath a calibrated burette or chemistry pipetting syringe containing the titrant. Small volumes of the titrant are then added to the analyte and indicator until the indicator changes color in reaction to the titrant saturation threshold, reflecting arrival at the endpoint of the titration. Depending on the endpoint desired, single drops or less than a single drop of the titrant can make the difference between a permanent and temporary change in the indicator. When the endpoint of the reaction is reached, the volume of reactant consumed is measured and used to calculate the concentration of analyte by

$$\mathbf{C}_a = \frac{\mathbf{C}_t \mathbf{V}_t \mathbf{M}}{\mathbf{V}_a}$$

where C_a is the concentration of the analyte, typically in molarity; C_t is the concentration of the titrant, typically in molarity; V_t is the volume of the titrant used, typically in liters; M is the mole ratio of the analyte and reactant from the balanced chemical equation; and V_a is the volume of the analyte used, typically in liters.

Preparation Techniques

Typical titrations require titrant and analyte to be in a liquid (solution) form. Though solids are usually dissolved into an aqueous solution, other solvents such as glacial acetic acid or ethanol are used for special purposes (as in petrochemistry). Concentrated analytes are often diluted to improve accuracy.

Many non-acid-base titrations require a constant pH throughout the reaction. Therefore a buffer solution may be added to the titration chamber to maintain the pH.

In instances where two reactants in a sample may react with the titrant and only one is the desired analyte, a separate masking solution may be added to the reaction chamber which masks the unwanted ion.

Some redox reactions may require heating the sample solution and titrating while the solution is still hot to increase the reaction rate. For instance, the oxidation of some oxalate solutions requires heating to 60 °C (140 °F) to maintain a reasonable rate of reaction.

Titration Curves

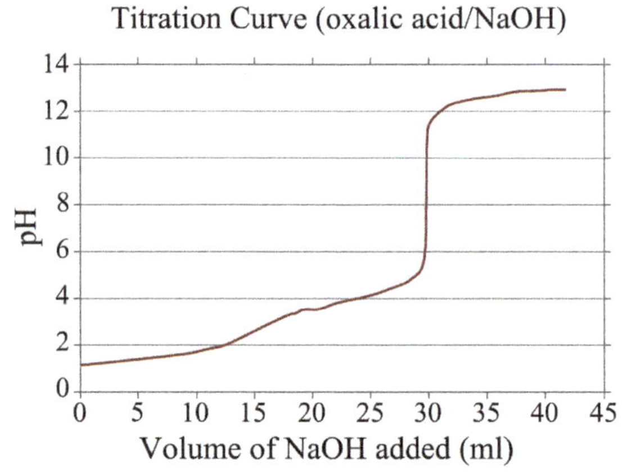

A typical titration curve of a diprotic acid titrated with a strong base. Shown here is oxalic acid titrated with sodium hydroxide. Both equivalence points are visible.

A titration curve is a curve in the plane whose x-coordinate is the volume of titrant added since the beginning of the titration, and whose y-coordinate is the concentration of the analyte at the corresponding stage of the titration (in an acid-base titration, the y-coordinate is usually the pH of the solution).

In an acid-base titration, the titration curve reflects the strength of the corresponding acid and base. For a strong acid and a strong base, the curve will be relatively smooth and very steep near the equivalence point. Because of this, a small change in titrant volume near the equivalence point results in a large pH change and many indicators would be appropriate (for instance litmus, phenolphthalein or bromothymol blue).

If one reagent is a weak acid or base and the other is a strong acid or base, the titration curve is irregular and the pH shifts less with small additions of titrant near the equivalence point. For example, the titration curve for the titration between oxalic acid (a weak acid) and sodium hydroxide (a strong base) is pictured. The equivalence point occurs between pH 8-10, indicating the solution is basic at the equivalence point and an indicator such as phenolphthalein would be appropriate. Titration curves corresponding to weak bases and strong acids are similarly behaved, with the solution being acidic at the equivalence point and indicators such as methyl orange and bromothymol blue being most appropriate.

Titrations between a weak acid and a weak base have titration curves which are highly irregular. Because of this, no definite indicator may be appropriate and a pH meter is often used to monitor the reaction.

The type of function that can be used to describe the curve is called a sigmoid function.

Types of Titrations

There are many types of titrations with different procedures and goals. The most common types of qualitative titration are acid-base titrations and redox titrations.

Acid–base Titration

Methyl orange

Indicator	Color on acidic side	Range of color change	Color on basic side
Methyl violet	Yellow	0.0–1.6	Violet
Bromophenol blue	Yellow	3.0–4.6	Blue

Methyl orange	Red	3.1–4.4	Yellow
Methyl red	Red	4.4–6.3	Yellow
Litmus	Red	5.0–8.0	Blue
Bromothymol blue	Yellow	6.0–7.6	Blue
Phenolphthalein	Colorless	8.3–10.0	Pink
Alizarin yellow	Yellow	10.1–12.0	Red

Acid-base titrations depend on the neutralization between an acid and a base when mixed in solution. In addition to the sample, an appropriate pH indicator is added to the titration chamber, reflecting the pH range of the equivalence point. The acid-base indicator indicates the endpoint of the titration by changing color. The endpoint and the equivalence point are not exactly the same because the equivalence point is determined by the stoichiometry of the reaction while the endpoint is just the color change from the indicator. Thus, a careful selection of the indicator will reduce the indicator error. For example, if the equivalence point is at a pH of 8.4, then the Phenolphthalein indicator would be used instead of Alizarin Yellow because phenolphthalein would reduce the indicator error. Common indicators, their colors, and the pH range in which they change color are given in the table above. When more precise results are required, or when the reagents are a weak acid and a weak base, a pH meter or a conductance meter are used.

For very strong bases, such as organolithium reagent, metal amides, and hydrides, water is generally not a suitable solvent and indicators whose pKa are in the range of aqueous pH changes are of little use. Instead, the titrant and indicator used are much weaker acids, and anhydrous solvents such as THF are used.

Redox Titration

Redox titrations are based on a reduction-oxidation reaction between an oxidizing agent and a reducing agent. A potentiometer or a redox indicator is usually used to determine the endpoint of the titration, as when one of the constituents is the oxidizing agent potassium dichromate. The color change of the solution from orange to green is not definite, therefore an indicator such as sodium diphenylamine is used. Analysis of wines for sulfur dioxide requires iodine as an oxidizing agent. In this case, starch is used as an indicator; a blue starch-iodine complex is formed in the presence of excess iodine, signalling the endpoint.

Some redox titrations do not require an indicator, due to the intense color of the constituents. For instance, in permanganometry a slight persisting pink color signals the endpoint of the titration because of the color of the excess oxidizing agent potassium permanganate. In iodometry, at sufficiently large concentrations, the disappearance of the deep red-brown triiodide ion can itself be used as an endpoint, though at lower concentrations sensitivity is improved by adding starch indicator, which forms an intensely blue complex with triiodide.

Color of iodometric titration mixture before (left) and after (right) the end point

Gas Phase Titration

Gas phase titrations are titrations done in the gas phase, specifically as methods for determining reactive species by reaction with an excess of some other gas, acting as the titrant. In one common gas phase titration, gaseous ozone is titrated with nitrogen oxide according to the reaction

$$O_3 + NO \rightarrow O_2 + NO_2.$$

After the reaction is complete, the remaining titrant and product are quantified (e.g., by FT-IR); this is used to determine the amount of analyte in the original sample.

Gas phase titration has several advantages over simple spectrophotometry. First, the measurement does not depend on path length, because the same path length is used for the measurement of both the excess titrant and the product. Second, the measurement does not depend on a linear change in absorbance as a function of analyte concentration as defined by the Beer-Lambert law. Third, it is useful for samples containing species which interfere at wavelengths typically used for the analyte.

Complexometric Titration

Complexometric titrations rely on the formation of a complex between the analyte and the titrant. In general, they require specialized complexometric indicators that form weak complexes with the analyte. The commonest example is the use of starch indicator to increase the sensitivity of iodometric titration, the dark blue complex of starch with iodine and iodide being more visible than iodine alone. Other complexometric indicators are Eriochrome Black T for the titration of calcium and magnesium ions, and the chelating agent EDTA used to titrate metal ions in solution.

Zeta Potential Titration

Zeta potential titrations are titrations in which the completion is monitored by the zeta potential, rather than by an indicator, in order to characterize heterogeneous systems, such as colloids. One of the uses is to determine the iso-electric point when surface charge becomes zero, achieved by changing the pH or adding surfactant. Another use is to determine the optimum dose for flocculation or stabilization.

Assay

An assay is a form of biological titration used to determine the concentration of a virus or bacterium. Serial dilutions are performed on a sample in a fixed ratio (such as 1:1, 1:2, 1:4, 1:8, etc.) until the last dilution does not give a positive test for the presence of the virus. The positive or negative value may be determined by visually inspecting the infected cells under a microscope or by an immunoenzymetric method such as enzyme-linked immunosorbent assay (ELISA). This value is known as the titer.

Measuring the Endpoint of a Titration

Different methods to determine the endpoint include:

- Indicator: A substance that changes color in response to a chemical change. An acid-base indicator (e.g., phenolphthalein) changes color depending on the pH. Redox indicators are also used. A drop of indicator solution is added to the titration at the beginning; the endpoint has been reached when the color changes.
- Potentiometer: An instrument that measures the electrode potential of the solution. These are used for redox titrations; the potential of the working electrode will suddenly change as the endpoint is reached.

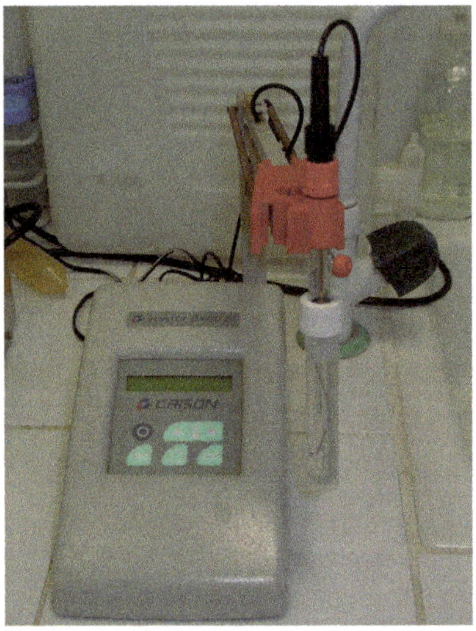

An elementary pH meter that can be used to monitor titration reactions

- pH meter: A potentiometer with an electrode whose potential depends on the amount of H^+ ion present in the solution. (This is an example of an ion-selective electrode.) The pH of the solution is measured throughout the titration, more accurately than with an indicator; at the endpoint there will be a sudden change in the measured pH.

- Conductivity: A measurement of ions in a solution. Ion concentration can change significantly in a titration, which changes the conductivity. (For instance, during an acid-base titration, the H^+ and OH^- ions react to form neutral H_2O.) As total conductance depends on all ions present in the solution and not all ions contribute equally (due to mobility and ionic strength), predicting the change in conductivity is more difficult than measuring it.

- Color change: In some reactions, the solution changes color without any added indicator. This is often seen in redox titrations when the different oxidation states of the product and reactant produce different colors.

- Precipitation: If a reaction produces a solid, a precipitate will form during the titration. A classic example is the reaction between Ag^+ and Cl^- to form the insoluble salt AgCl. Cloudy precipitates usually make it difficult to determine the endpoint precisely. To compensate, precipitation titrations often have to be done as "back" titrations.

- Isothermal titration calorimeter: An instrument that measures the heat produced or consumed by the reaction to determine the endpoint. Used in biochemical titrations, such as the determination of how substrates bind to enzymes.

- Thermometric titrimetry: Differentiated from calorimetric titrimetry because the heat of the reaction (as indicated by temperature rise or fall) is not used to determine the amount of analyte in the sample solution. Instead, the endpoint is determined by *the rate of temperature change.*

- Spectroscopy: Used to measure the absorption of light by the solution during titration if the spectrum of the reactant, titrant or product is known. The concentration of the material can be determined by Beer's Law.

- Amperometry: Measures the current produced by the titration reaction as a result of the oxidation or reduction of the analyte. The endpoint is detected as a change in the current. This method is most useful when the excess titrant can be reduced, as in the titration of halides with Ag^+.

Endpoint and Equivalence Point

Though equivalence point and endpoint are used interchangeably, they are different terms. *Equivalence point* is the theoretical completion of the reaction: the volume of added titrant at which the number of moles of titrant is equal to the number of moles of analyte, or some multiple thereof (as in polyprotic acids). *Endpoint* is what is actually measured, a physical change in the solution as determined by an indicator or an instrument mentioned above.

There is a slight difference between the endpoint and the equivalence point of the titration. This error is referred to as an indicator error, and it is indeterminate.

Back Titration

Back titration is a titration done in reverse; instead of titrating the original sample, a known excess of standard reagent is added to the solution, and the excess is titrated. A back titration is useful if the endpoint of the reverse titration is easier to identify than the endpoint of the normal titration, as with precipitation reactions. Back titrations are also useful if the reaction between the analyte and the titrant is very slow, or when the analyte is in a non-soluble solid.

Graphical Methods

The titration process creates solutions with compositions ranging from pure acid to pure base. Identifying the pH associated with any stage in the titration process is relatively simple for monoprotic acids and bases. The presence of more than one acid or base group complicates these computations. Graphical methods, such as the equiligraph, have long been used to account for the interaction of coupled equilibria. These graphical solution methods are simple to implement, however they are infrequently used.

Particular Uses

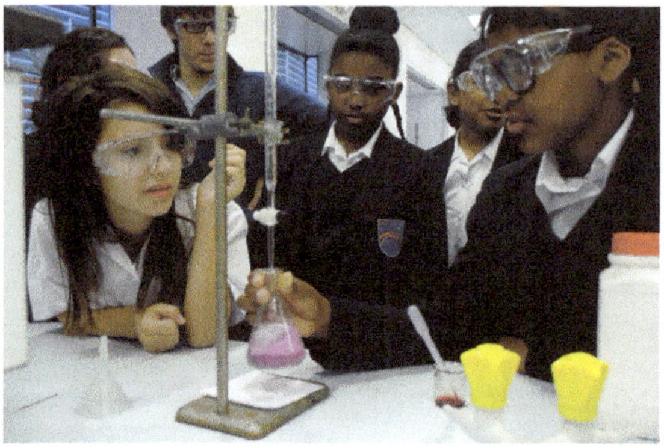

A titration is demonstrated to secondary school students.

Acid-base Titrations

- In biodiesel: Waste vegetable oil (WVO) must be neutralized before a batch may be processed. A portion of WVO is titrated with a base to determine acidity, so the rest of the batch may be properly neutralized. This removes free fatty acids from the WVO that would normally react to make soap instead of biodiesel.
- Kjeldahl method: A measure of nitrogen content in a sample. Organic nitrogen is digested into ammonia with sulfuric acid and potassium sulfate. Finally, ammonia is back titrated with boric acid and then sodium carbonate.

- Acid value: The mass in milligrams of potassium hydroxide (KOH) required to titrate fully an acid in one gram of sample. An example is the determination of free fatty acid content.
- Saponification value: The mass in milligrams of KOH required to saponify a fatty acid in one gram of sample. Saponification is used to determine average chain length of fatty acids in fat.
- Ester value (or ester index): A calculated index. Ester value = Saponification value – Acid value.
- Amine value: The mass in milligrams of KOH equal to the amine content in one gram of sample.
- Hydroxyl value: The mass in milligrams of KOH corresponding to hydroxyl groups in one gram of sample. The analyte is acetylated using acetic anhydride then titrated with KOH.

Redox Titrations

- Winkler test for dissolved oxygen: Used to determine oxygen concentration in water. Oxygen in water samples is reduced using manganese(II) sulfate, which reacts with potassium iodide to produce iodine. The iodine is released in proportion to the oxygen in the sample, thus the oxygen concentration is determined with a redox titration of iodine with thiosulfate using a starch indicator.
- Vitamin C: Also known as ascorbic acid, vitamin C is a powerful reducing agent. Its concentration can easily be identified when titrated with the blue dye Dichlorophenolindophenol (DCPIP) which turns colorless when reduced by the vitamin.
- Benedict's reagent: Excess glucose in urine may indicate diabetes in the patient. Benedict's method is the conventional method to quantify glucose in urine using a prepared reagent. In this titration, glucose reduces cupric ions to cuprous ions which react with potassium thiocyanate to produce a white precipitate, indicating the endpoint.
- Bromine number: A measure of unsaturation in an analyte, expressed in milligrams of bromine absorbed by 100 grams of sample.
- Iodine number: A measure of unsaturation in an analyte, expressed in grams of iodine absorbed by 100 grams of sample.

Miscellaneous

- Karl Fischer titration: A potentiometric method to analyze trace amounts of water in a substance. A sample is dissolved in methanol, and titrated with Karl Fischer reagent. The reagent contains iodine, which reacts proportionally with water. Thus, the water content can be determined by monitoring the potential of excess iodine.

Tallow

Tallow is a rendered form of beef or mutton fat, processed from suet. It is solid at room temperature. Unlike suet, tallow can be stored for extended periods without the need for refrigeration to prevent decomposition, provided it is kept in an airtight container to prevent oxidation.

Tallow made by rendering calf suet

In industry, tallow is not strictly defined as beef or mutton fat. In this context, tallow is animal fat that conforms to certain technical criteria, including its melting point. It is common for commercial tallow to contain fat derived from other animals, such as lard from pigs, or even from plant sources.

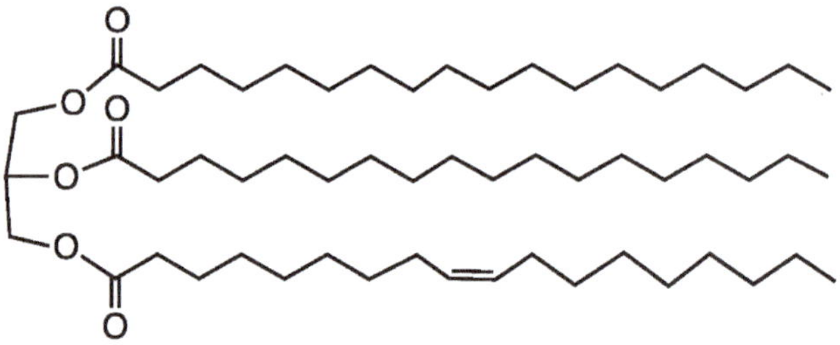

Tallow consists mainly of triglycerides (fat), whose major constituents are derived from stearic and oleic acids.

The diagram to the right shows the chemical structure of a typical triglyceride molecule.

Uses

Tallow is used mainly in producing soap and animal feed.

Production of Biodiesel

Tallow can be used for the production of biodiesel in much the same way as oils from plants are currently used. Because tallow is derived from animal by-products which have little to no value to commercial food industries, it avoids some of the food vs. fuel debate.

Aviation Fuel

The United States Air Force has experimented successfully with the use of beef tallow in aviation biofuels. During five days of flight testing from August 23 to 27, 2010, at Edwards Air Force Base, California, a U.S. Air Force C-17 Globemaster III flew using JP-8 conventional jet fuel in three of its engines and a 50/50 blend of JP-8 and HRJ biofuel made from beef tallow in one engine on 23 August, followed by a flight with the same 50/50 blend in all four engines on 24 August. On 27 August, it flew using a blend of 50 percent JP-8, 25 percent HRJ, and 25 percent coal-based fuel made through the Fischer–Tropsch process, becoming the first United States Department of Defense aircraft to fly on such a blend and the first aircraft to operate from Edwards using a fuel derived from beef tallow.

Food

A significant use of tallow is for the production of shortening. It is one of the main ingredients of Native American food called pemmican. Tallow is sometimes used in deep frying in place of other oils. Before switching to pure vegetable oil in 1990, the McDonald's corporation cooked its French fries in a mixture of 93% beef tallow and 7% cottonseed oil. Currently, McDonald's French fries and hash browns contain beef flavoring derived from hydrolyzed wheat and milk that is added during production.

Niche Uses

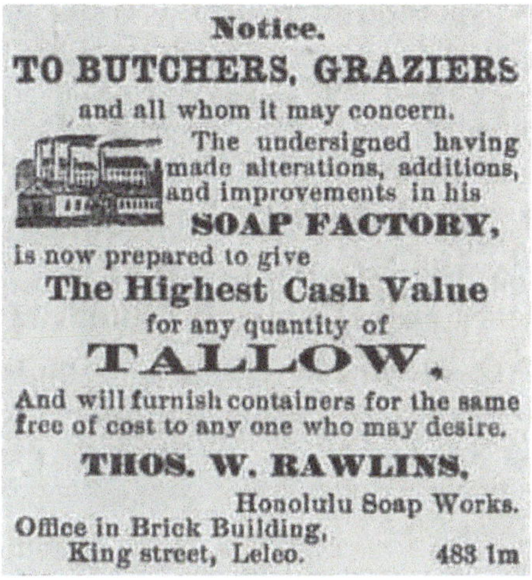

An 1883 ad soliciting tallow from butchers and graziers for soap production in the Hawaiʻi newspaper The Daily Bulletin

Many items of traditional goods are produced from tallow, which was widely available domestically. Tallow can also be used as flux for soldering. It is also the primary ingredient in some leather conditioners. Tallow used to be used commonly in high-end shaving soaps, in particular those of elite British firms such as Geo. F Trumper, Truefitt & Hill, and Taylor of Old Bond Street. While these firms have reformulated to a vegetable base, tallow-based soaps still exist, including the soaps from American firms, Williams Mug Shaving Soap, Nod Hill Soap, and soaps from Turkey and Italy. Tallow-based shaving soaps have enjoyed a resurgence in recent years due to gaining popularity of traditional wet-shaving, with many artisan and non-artisan makers using tallow in their formula.

Tallow also has a use in printmaking where it is combined with bitumen and applied to metal print plates to provide a resist to acid etching.

Historic Applications

Biblical Mention

In Leviticus 3:14-17 the Israelites are forbidden to eat the suet surrounding certain internal organs of animals (e.g. cows and sheep) sacrificed at the Temple. This suet is Halakhically called *chelev*. English Bible translations sometimes translate *chelev* to "tallow", although strictly speaking this is incorrect: only tallow from species offered for sacrifice at the Temple is forbidden, and tallow from other (typically wild) kosher quadrupeds (e.g. deer) is not forbidden.

Candles

Tallow once was widely used to make moulded candles before more convenient wax varieties became available—and for some time after, as they continued to be a cheaper alternative. For those too poor even to avail themselves of homemade, moulded tallow candles, the "tallow dip"—a strip of burning cloth in a saucer of tallow grease—was an accessible substitute.

Lubrication

Early in the development of steam-driven piston engines, the hot vapors and liquids washed away most lubricants very quickly. It was soon found that tallow was quite resistant to this washing. Tallow and compounds including tallow were widely used to lubricate locomotive and steamship engines at least until the 1950s. (During WWII, the vast fleets of steam-powered ships exhausted the supply, leading to the large-scale planting of rapeseed because rapeseed oil also resisted the washing effect.) Tallow is still used in the steel rolling industry to provide the required lubrication as the sheet steel is compressed through the steel rollers. There is a trend toward replacing tallow-based lubrication with synthetic oils in rolling applications for surface cleanliness reasons.

The use of tallow or lard to lubricate rifles was the spark that started the Indian Mutiny of 1857. To load the new Pattern 1853 Enfield Rifle, the sepoys had to bite the cartridge open. It was believed that the paper cartridges that were standard issue with the rifle were greased with lard (pork fat), which was regarded as unclean by Muslims, or tallow (cow fat), which is incompatible with Hindu dietary laws. Tallow, along with beeswax, was also used in the lubricant for American Civil War ammunition used in the Springfield rifled musket. A combination of mutton tallow, paraffin wax and beeswax is still used as a patch or projectile lubricant in present-day black powder arms.

Tallow is used to make a biodegradable motor oil by a Stamford, Connecticut-based company called Green Earth Technologies.

Tallow finds a number of uses in woodworking. It can serve as a lubricant and rust inhibitor on tools such as saw blades and cast-iron plane bodies. It can also be helpful with screw fixings into hard woods, particularly when brass screws are employed. It reduces the risk of screw breakage during initial screw penetration and also eases later screw removal should this be required, by preventing corrosion.

Medicinal

In Germany, deer tallow (*Hirschtalg*) is used as a base ingredient in certain salves used by athletes to prevent sore skins or blisters.

Composition

The composition of the fatty acids is typically as follows:

- Saturated fatty acids:
    - Palmitic acid (C16:0): 26%
    - Stearic acid (C18:0): 14%
    - Myristic acid (C14:0): 3%
- Monounsaturated fatty acids:
    - Oleic acid (C18-1, ω-9): 47%
    - Palmitoleic acid (C16:1)]: 3%
- Polyunsaturated fatty acids:
    - Linoleic acid: 3%
    - Linolenic acid: 1%

Fatty Acid Methyl Ester

Fatty acid methyl esters (FAME) are a type of fatty acid ester that are derived by transesterification of fats with methanol. The molecules in biodiesel are primarily FAMEs, usually obtained from vegetable oils by transesterification. They are used to produce detergents and biodiesel. FAMES are typically produced by an alkali-catalyzed reaction between fats and methanol in the presence of base such as sodium hydroxide or sodium methoxide.

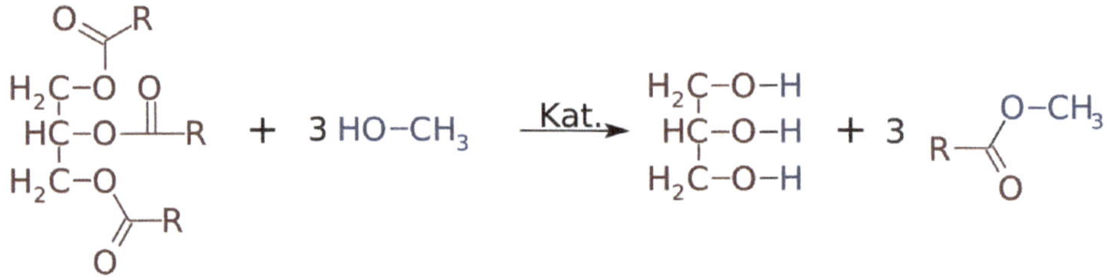

Other Details

Since every microorganism has its specific FAME profile (microbial fingerprinting), it can be used as a tool for microbial source tracking (MST). The types and proportions of fatty acids present in cytoplasm membrane and outer membrance (gram negative) lipids of cells are major phenotypic traits.

Clinical analysis can determine the lengths, bonds, rings and branches of the FAME. To perform this analysis, a bacterial culture is taken, and the fatty acids extracted and used to form methyl esters. The volatile derivatives are then introduced into a gas chromatagraph, and the patterns of the peaks help identify the organism. This is widely used in characterizing new species of bacteria, and is useful for identifying pathogenic strains.

Husk

Corn being husked in the yard of a Dungan farmer in Kyrgyzstan

Husk (or hull) in botany is the outer shell or coating of a seed. It often refers to the leafy outer covering of an ear of maize (corn) as it grows on the plant. Literally, a husk or hull includes the protective outer covering of a seed, fruit or vegetable. It can also refer to the exuvia of bugs or small animals left behind after moulting.

In cooking, hull can also refer to other waste parts of fruits and vegetables, notably the cap or sepal of a strawberry.

The husk of a legume and some similar fruits is called a pod.

Plantago-seed mucilage is often referred to as husk, or psyllium husk.

Crop plants of several species have been selected that have hulless seeds, including pumpkins, oats, and barley.

Women in Cape Verde using multiple pestles in a large mortar

Husking and Dehulling

Husking of corn is the process of removing its outer layers, leaving only the cob or seed rack of the corn. Dehulling is the process of removing the hulls (or chaff) from beans and other seeds. This is sometimes done using a machine known as a huller. To prepare the seeds to have oils extracted from them, they are cleaned to remove any foreign objects. Next, the seeds have their hulls, or outer coverings, or husk, removed. There are three different types of dehulling systems that can be used to process soybeans: Hot dehulling, warm dehulling and cold dehulling. Hot dehulling is the system offered in areas where beans are processed directly from the field. Warm dehulling is often used by processors who import their soybeans. Cold dehulling is used in plants that have existing drying and conditioning equipment, but need to add dehulling equipment to produce high protein meal. The different dehulling temperature options are for different types of production, beans and preparation equipment.

In third-world countries, husking and dehulling is still often done by hand using a large mortar and pestle. These are usually made of wood, and operated by one or more people.

The husk is biodegradable and may be composted.

References

- Harris, D.C. (2007). Quantitative Chemical Analysis (7 ed.). W. H. Freeman and Company. p. 12. ISBN 9780716770411.
- Matar, S.; L.F. Hatch (2001). Chemistry of Petrochemical Processes (2 ed.). Gulf Professional Publishing. ISBN 0-88415-315-0.
- Verma, Dr. N.K.; S.K. Khanna; Dr B. Kapila. Comprehensive Chemistry XI. New Dehli: Laxmi Publications. pp. 642–645. ISBN 81-7008-596-9.
- Patnaik, P. (2004). Dean's Analytical Chemistry Handbook (2 ed.). McGraw-Hill Prof Med/Tech. pp. 2.11–2.16. ISBN 0-07-141060-0.
- Reger, D.L.; S.R. Goode; D.W. Ball (2009). Chemistry: Principles and Practice (3 ed.). Cengage Learning. pp. 684–693. ISBN 0-534-42012-5.
- Vogel, A.I.; J. Mendham (2000). Vogel's textbook of quantitative chemical analysis (6 ed.). Prentice Hall. p. 423. ISBN 0-582-22628-7.
- Amerine, M.A.; M.A. Joslyn (1970). Table wines: the technology of their production. 2 (2 ed.). University of California Press. pp. 751–753. ISBN 0-520-01657-2.
- Khopkar, S.M. (1998). Basic Concepts of Analytical Chemistry (2 ed.). New Age International. pp. 63–76. ISBN 81-224-1159-2.
- Somasundaran, P. (2006). "Calculation of Zeta-Potentials from Electrokinetic Data". Encyclopedia of Surface and Colloid Science (2 ed.). CRC Press. 2: 1097. ISBN 0-8493-9607-7.
- Decker, J.M. (2000). Introduction to immunology. Eleventh Hour (3 ed.). Wiley-Blackwell. pp. 18–20. ISBN 0-632-04415-2.
- Hannan, H.J. (2007). Technician's Formulation Handbook for Industrial and Household Cleaning Products. Lulu.com. p. 103. ISBN 0-615-15601-0.

5

Transesterification: A Comprehensive Study

The procedure of replacing organic group R" of an ester with organic R' of an alcohol is known as transesterification. The common transesterification catalysts explained in this section are sodium hydroxide, potassium hydroxide and sodium methoxide. This text has been carefully written to provide an easy understanding of the varied facets of transesterification.

Transesterification

In organic chemistry, transesterification is the process of exchanging the organic group R″ of an ester with the organic group R′ of an alcohol. These reactions are often catalyzed by the addition of an acid or base catalyst. The reaction can also be accomplished with the help of enzymes (biocatalysts) particularly lipases (E.C.3.1.1.3).

Transesterification: alcohol + ester → different alcohol + different ester

Strong acids catalyse the reaction by donating a proton to the carbonyl group, thus making it a more potent electrophile, whereas bases catalyse the reaction by removing a proton from the alcohol, thus making it more nucleophilic. Esters with larger alkoxy groups can be made from methyl or ethyl esters in high purity by heating the mixture of ester, acid/base, and large alcohol and evaporating the small alcohol to drive equilibrium.

Mechanism

In the transesterification mechanism, the carbonyl carbon of the starting ester ($RCOOR^1$) undergoes nucleophilic attack by the incoming alkoxide (R^2O^-) to give a tetrahedral intermediate, which either reverts to the starting material, or proceeds to the transesterified product ($RCOOR^2$). The various species exist in equilibrium, and the product distribution depends on the relative energies of the reactant and product.

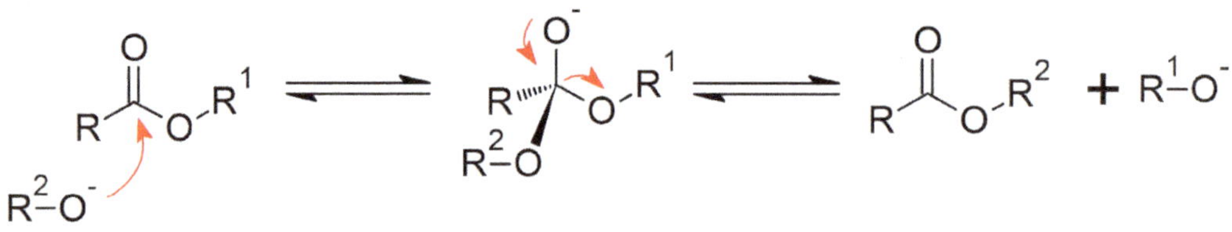

Applications

Polyester Production

The largest scale application of transesterification is in the synthesis of polyesters. In this application diesters undergo transesterification with diols to form macromolecules. For example, dimethyl terephthalate and ethylene glycol react to form polyethylene terephthalate and methanol, which is evaporated to drive the reaction forward.

Methanolysis and Biodiesel Production

The reverse reaction, methanolysis, is also an example of transesterification. This process has been used to recycle polyesters into individual monomers (see plastic recycling). It is also used to convert fats (triglycerides) into biodiesel. This conversion was one of the first uses. Transesterified vegetable oil (biodiesel) was used to power heavy-duty vehicles in South Africa before World War II.

It was patented in the US in the 1950s by Colgate, though biolipid transesterification may have been discovered much earlier. In the 1940s, researchers were looking for a method to more readily produce glycerol, which was used to produce explosives for World War II. Many of the methods used today by producers and home brewers have their origin in the original 1940s research.

Biolipid transesterification has also been recently shown by Japanese researchers to be possible using a super-critical methanol methodology, whereby high temperature, high-pressure vessels are used to physically catalyze the biolipid/methanol reaction into fatty-acid methyl esters.

High-pressure Transesterification

Base-catalyzed transesterification is characterized by a negative activation volume (approx. −12 cm^3/mol) and therefore it proceeds faster under high-pressure conditions. It has been shown that amine-catalyzed alcoholysis of sterically hindered esters (e.g. protecting groups, chiral auxiliaries) proceeds rapidly at room temperature under 10 kbar pressure, giving quantitative yields.

Common Transesterification Catalyst

Sodium Hydroxide

Sodium hydroxide (NaOH), also known as lye and caustic soda, is an inorganic compound. It is a white solid and highly caustic metallic base and alkali of sodium which is available in pellets, flakes, granules, and as prepared solutions at different concentrations. Sodium hydroxide forms an approximately 50% (by mass) saturated solution with water.

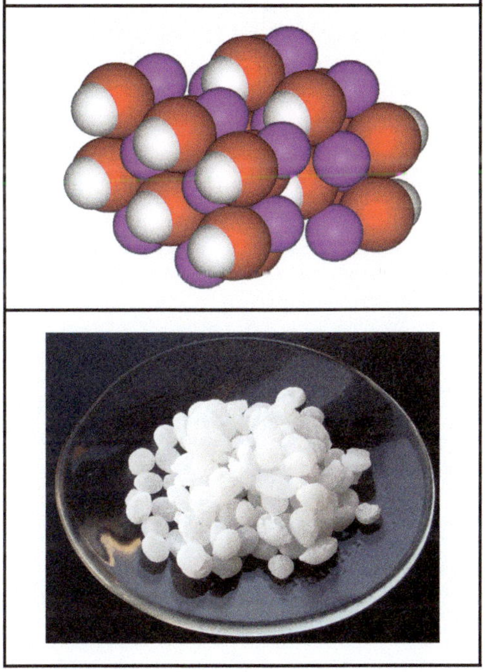

Sodium hydroxide

Sodium hydroxide is soluble in water, ethanol, and methanol. This alkali is deliquescent and readily absorbs moisture and carbon dioxide in air.

Sodium hydroxide is used in many industries, mostly as a strong chemical base in the manufacture of pulp and paper, textiles, drinking water, soaps and detergents and as a drain cleaner. Worldwide production in 2004 was approximately 60 million tonnes, while demand was 51 million tonnes.

Properties

Physical Properties

Pure sodium hydroxide is a whitish solid, sold in pellets, flakes, and granular form, as well as in solution. It is highly soluble in water, with a lower solubility in ethanol and methanol, but is insoluble in ether and other non-polar solvents.

Similar to the hydration of sulfuric acid, dissolution of solid sodium hydroxide in water is a highly exothermic reaction in which a large amount of heat is liberated, posing a threat to safety through the possibility of splashing. The resulting solution is usually colourless and odorless. As with other alkaline solutions, it feels slippery when it comes in contact with skin.

Chemical Properties

Reaction with Acids

Sodium hydroxide reacts with protic acids to produce water and the corresponding salts. For example, when sodium hydroxide reacts with hydrochloric acid, sodium chloride is formed:

$$NaOH(aq) + HCl(aq) \rightarrow NaCl(aq) + H_2O(l)$$

In general, such neutralization reactions are represented by one simple net ionic equation:

$$OH^-(aq) + H^+(aq) \rightarrow H_2O(l)$$

This type of reaction with a strong acid releases heat, and hence is exothermic. Such acid-base reactions can also be used for titrations. However, sodium hydroxide is not used as a primary standard because it is hygroscopic and absorbs carbon dioxide from air.

Reaction with Acidic Oxides

Sodium hydroxide also reacts with acidic oxides, such as sulfur dioxide. Such reactions are often used to "scrub" harmful acidic gases (like SO_2 and H_2S) produced in the burning of coal and thus prevent their release into the atmosphere. For example,

$$2\ NaOH + SO_2 \rightarrow Na_2SO_3 + H_2O$$

Reaction with Amphoteric Metals and Oxides

Glass reacts slowly with aqueous sodium hydroxide solutions at ambient temperatures to form soluble silicates. Because of this, glass joints and stopcocks exposed to sodium hydroxide have a tendency to "freeze". Flasks and glass-lined chemical reactors are damaged by long exposure to hot sodium hydroxide, which also frosts the glass. Sodium hydroxide does not attack iron since iron does not have amphoteric properties (i.e., it only dissolves in acid, not base). A few transition metals, however, may react vigorously with sodium hydroxide.

In 1986, an aluminium road tanker in the UK was mistakenly used to transport 25% sodium hydroxide solution, causing pressurization of the contents and damage to the tanker. The pressurization was due to the hydrogen gas which is produced in the reaction between sodium hydroxide and aluminium:

$$2\,Al + 2\,NaOH + 6\,H_2O \rightarrow 2\,NaAl(OH)_4 + 3\,H_2$$

Precipitant

Unlike sodium hydroxide, the hydroxides of most transition metals are insoluble, and therefore sodium hydroxide can be used to precipitate transition metal hydroxides. The following colours are observed: blue-copper, green-iron(II), yellow/brown-iron(III). Zinc and lead salts dissolve in excess sodium hydroxide to give a clear solution of Na_2ZnO_2 or Na_2PbO_2.

Aluminium hydroxide is used as a gelatinous flocculant to filter out particulate matter in water treatment. Aluminium hydroxide is prepared at the treatment plant from aluminium sulfate by reacting it with sodium hydroxide or carbonate.

$$Al_2(SO_4)_3 + 6\,NaOH \rightarrow 2\,Al(OH)_3 + 3\,Na_2SO_4$$

$$Al_2(SO_4)_3 + 6\,NaHCO_3 \rightarrow 2\,Al(OH)_3 + 3\,Na_2SO_4 + 6\,CO_2$$

Saponification

Sodium hydroxide can be used for the base-driven hydrolysis of esters (as in saponification), amides and alkyl halides. However, the limited solubility of sodium hydroxide in organic solvents means that the more soluble potassium hydroxide (KOH) is often preferred. Touching sodium hydroxide solution with the bare hands, while not recommended, produces a slippery feeling. This happens because the oils of the hand are converted to soap.

Production

Sodium hydroxide is industrially produced as a 50% solution by variations of the electrolytic chloralkali process. Chlorine gas is also produced in this process. Solid sodium hydroxide is obtained from this solution by the evaporation of water. Solid sodium hydroxide is most commonly sold as flakes, prills, and cast blocks.

In 2004, world production was estimated at 60 million dry metric tonnes of sodium hydroxide, and demand was estimated at 51 million tonnes. In 1998, total world production was around 45 million tonnes. North America and Asia each contributed around 14 million tonnes, while Europe produced around 10 million tonnes. In the United States, the major producer of sodium hydroxide is the Dow Chemical Company, which has annual production around 3.7 million tonnes from sites at Freeport, Texas, and Plaquemine, Louisiana. Other major US producers include Oxychem, PPG, Olin, Pioneer Companies, Inc. (PIONA, which was purchased by Olin), and Formosa. All of these companies use the chloralkali process.

Historically, sodium hydroxide was produced by treating sodium carbonate with calcium hydroxide in a metathesis reaction. (Sodium hydroxide is soluble while calcium carbonate is not.) This process was called causticizing.

$$Ca(OH)_2(aq) + Na_2CO_3(s) \rightarrow CaCO_3\downarrow + 2\ NaOH(aq)$$

This process was superseded by the Solvay process in the late 19th century, which was in turn supplanted by the chloralkali process which we use today.

Sodium hydroxide is also produced by combining pure sodium metal with water. The byproducts are hydrogen gas and heat, often resulting in a flame, making this a common demonstration of the reactivity of alkali metals in academic environments; however, it is not commercially viable, as the isolation of sodium metal is typically performed by reduction or electrolysis of sodium compounds including sodium hydroxide.

Uses

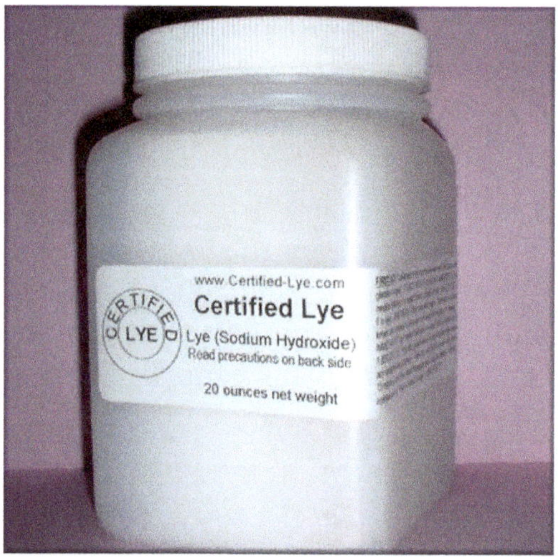

Canister of sodium hydroxide.

Sodium hydroxide is a popular strong base used in industry. Around 56% of sodium hydroxide produced is used by industry, 25% of which is used in the paper industry. Sodium hydroxide is also used in the manufacture of sodium salts and detergents, pH regulation, and organic synthesis. It is used in the Bayer process of aluminium production. In bulk, it is most often handled as an aqueous solution, since solutions are cheaper and easier to handle.

Sodium hydroxide is used in many scenarios where it is desirable to increase the alkalinity of a mixture, or to neutralize acids.

For example, in the petroleum industry, sodium hydroxide is used as an additive in drilling mud to increase alkalinity in bentonite mud systems, to increase the mud viscosity, and to neutralize any acid gas (such as hydrogen sulfide and carbon dioxide) which may be encountered in the geological formation as drilling progresses.

Poor quality crude oil can be treated with sodium hydroxide to remove sulfurous impurities in a process known as *caustic washing*. As above, sodium hydroxide reacts with weak acids such as hydrogen sulfide and mercaptans to yield non-volatile sodium salts, which can be removed. The waste which is formed is toxic and difficult to deal with, and the process is banned in many countries because of this. In 2006, Trafigura used the process and then dumped the waste in Africa.

Chemical Pulping

Sodium hydroxide is also widely used in pulping of wood for making paper or regenerated fibers. Along with sodium sulfide, sodium hydroxide is a key component of the white liquor solution used to separate lignin from cellulose fibers in the kraft process. It also plays a key role in several later stages of the process of bleaching the brown pulp resulting from the pulping process. These stages include oxygen delignification, oxidative extraction, and simple extraction, all of which require a strong alkaline environment with a pH > 10.5 at the end of the stages.

Tissue Digestion

In a similar fashion, sodium hydroxide is used to digest tissues, as in a process that was used with farm animals at one time. This process involved placing a carcass into a sealed chamber, then adding a mixture of sodium hydroxide and water (which breaks the chemical bonds that keep the flesh intact). This eventually turns the body into a liquid with coffee-like appearance, and the only solid that remains are bone hulls, which could be crushed between one's fingertips. Sodium hydroxide is frequently used in the process of decomposing roadkill dumped in landfills by animal disposal contractors. Due to its low cost and availability, it has been used to dispose of corpses by criminals. Italian serial killer Leonarda Cianciulli used this chemical to turn dead bodies into soap. In Mexico, a man who worked for drug cartels admitted disposing of over 300 bodies with it. Sodium hydroxide is a dangerous chemical due to its ability to hydrolyze protein. If a dilute solution is spilled on the skin, burns may result if the area is not washed thoroughly and for several minutes with running water. Splashes in the eye can be more serious and can lead to blindness. This danger is often overlooked.

Dissolving Amphoteric Metals and Compounds

Strong bases attack aluminium. Sodium hydroxide reacts with aluminium and water to release hydrogen gas. The aluminium takes the oxygen atom from sodium hydroxide, which in turn takes the oxygen atom from the water, and releases the two hydrogen atoms, The reaction thus produces hydrogen gas and sodium aluminate. In this reaction, sodium hydroxide acts as an agent to make the solution alkaline, which aluminium can dissolve in. This reaction can be useful in etching, removing anodizing, or converting a polished surface to a satin-like finish, but without further passivation such as anodizing or alodining the surface may become degraded, either under normal use or in severe atmospheric conditions.

In the Bayer process, sodium hydroxide is used in the refining of alumina containing ores (bauxite) to produce alumina (aluminium oxide) which is the raw material used to produce aluminium metal via the electrolytic Hall-Héroult process. Since the alumina is amphoteric, it dissolves in the sodium hydroxide, leaving impurities less soluble at high pH such as iron oxides behind in the form of a highly alkaline red mud.

Other amphoteric metals are zinc and lead which dissolve in concentrated sodium hydroxide solutions to give sodium zincate and sodium plumbate respectively.

Esterification and Transesterification Reagent

Sodium hydroxide is traditionally used in soap making (cold process soap, saponification). It was made in the nineteenth century for a hard surface rather than liquid product because it was easier to store and transport.

For the manufacture of biodiesel, sodium hydroxide is used as a catalyst for the transesterification of methanol and triglycerides. This only works with anhydrous sodium hydroxide, because combined with water the fat would turn into soap, which would be tainted with methanol. NaOH is used more often than potassium hydroxide because it is cheaper and a smaller quantity is needed.

Sodium hydroxide is also being used experimentally in a new technology to create synthetic gasoline.

Food Preparation

Food uses of sodium hydroxide include washing or chemical peeling of fruits and vegetables, chocolate and cocoa processing, caramel coloring production, poultry scalding, soft drink processing, and thickening ice cream. Olives are often soaked in sodium hydroxide for softening; Pretzels and German lye rolls are glazed with a sodium hydroxide solution before baking to make them crisp. Owing to the difficulty in obtaining food grade sodium hydroxide in small quantities for home use, sodium carbonate is often used in place of sodium hydroxide.

Specific foods processed with sodium hydroxide include:

- The Scandinavian delicacy known as lutefisk (from *lutfisk*, "lye fish").
- Hominy is dried maize (corn) kernels reconstituted by soaking in lye-water. These expand considerably in size and may be further processed by frying to

make corn nuts or by drying and grinding to make grits. Nixtamal is similar, but uses calcium hydroxide instead of sodium hydroxide.

- Sodium hydroxide is also the chemical that causes gelling of egg whites in the production of Century eggs.
- German pretzels are poached in a boiling sodium carbonate solution or cold sodium hydroxide solution before baking, which contributes to their unique crust.
- Lye-water is an essential ingredient in the crust of the traditional baked Chinese moon cakes.
- Most yellow coloured Chinese noodles are made with lye-water but are commonly mistaken for containing egg.
- Some methods of preparing olives involve subjecting them to a lye-based brine.
- The Filipino dessert (*kakanin*) called *kutsinta* uses a small quantity of lye water to help give the rice flour batter a jelly like consistency. A similar process is also used in the kakanin known as *pitsi-pitsi* or *pichi-pichi* except that the mixture uses grated cassava instead of rice flour.

Cleaning Agent

Sodium hydroxide is frequently used as an industrial cleaning agent where it is often called "caustic". It is added to water, heated, and then used to clean process equipment, storage tanks, etc. It can dissolve grease, oils, fats and protein-based deposits. It is also used for cleaning waste discharge pipes under sinks and drains in domestic properties. Surfactants can be added to the sodium hydroxide solution in order to stabilize dissolved substances and thus prevent redeposition. A sodium hydroxide soak solution is used as a powerful degreaser on stainless steel and glass bakeware. It is also a common ingredient in oven cleaners.

A common use of sodium hydroxide is in the production of parts washer detergents. Parts washer detergents based on sodium hydroxide are some of the most aggressive parts washer cleaning chemicals. The sodium hydroxide-based detergents include surfactants, rust inhibitors and defoamers. A parts washer heats water and the detergent in a closed cabinet and then sprays the heated sodium hydroxide and hot water at pressure against dirty parts for degreasing applications. Sodium hydroxide used in this manner replaced many solvent-based systems in the early 1990s when trichloroethane was outlawed by the Montreal Protocol. Water and sodium hydroxide detergent-based parts washers are considered to be an environmental improvement over the solvent-based cleaning methods.

Sodium hydroxide is used in the home as a type of drain opener to unblock clogged drains, usually in the form of a dry crystal or as a thick liquid gel. The alkali dissolves greases to produce water soluble products. It also hydrolyzes the proteins such as those found in hair which may block water pipes. These reactions are sped by the heat generated when sodium hydroxide and the other chemical components of the cleaner dissolve in water. Such alkaline drain cleaners and their acidic versions are highly corrosive and should be handled with great caution.

Hardware stores grade sodium hydroxide to be used as a type of drain cleaner.

Sodium hydroxide is used in some relaxers to straighten hair. However, because of the high incidence and intensity of chemical burns, manufacturers of chemical relaxers use other alkaline chemicals in preparations available to average consumers. Sodium hydroxide relaxers are still available, but they are used mostly by professionals.

Paint stripping with caustic soda

A solution of sodium hydroxide in water was traditionally used as the most common paint stripper on wooden objects. Its use has become less common, because it can damage the wood surface, raising the grain and staining the colour.

Water Treatment

Sodium hydroxide is sometimes used during water purification to raise the pH of water supplies. Increased pH makes the water less corrosive to plumbing and reduces the amount of lead, copper and other toxic metals that can dissolve into drinking water.

Historical Uses

Sodium hydroxide has been used for detection of carbon monoxide poisoning, with blood samples of such patients turning to a vermilion color upon the addition of a few drops of sodium hydroxide. Today, carbon monoxide poisoning can be detected by CO oximetry.

Experimental

- Sodium hydroxide test for flavonoids

Safety

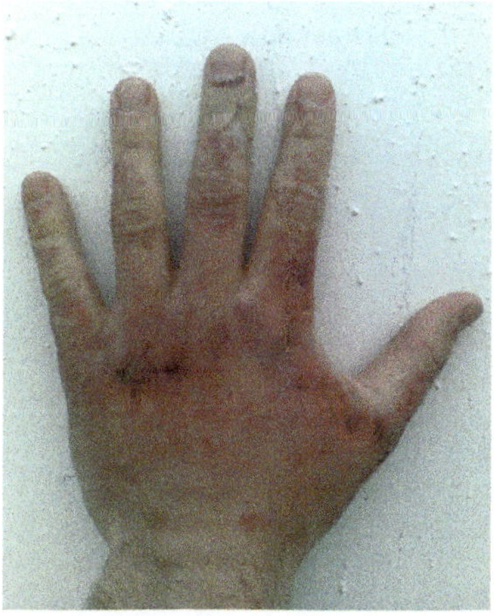

Chemical burns caused by sodium hydroxide solution photographed 44 hours after exposure.

Like other corrosive acids and alkalis, drops of sodium hydroxide solutions can readily decompose proteins and lipids in living tissues via amide hydrolysis and ester hydrolysis, which consequently cause chemical burns and may induce permanent blindness upon contact with eyes. Solid alkali can also express its corrosive nature if there is water, such as water vapor. Thus, protective equipment, like rubber gloves, safety clothing and eye protection, should always be used when handling this chemical or its solutions. The standard first aid measures for alkali spills on the skin is, as for other corrosives, irrigation with large quantities of water. Washing is continued for at least ten to fifteen minutes.

Moreover, dissolution of sodium hydroxide is highly exothermic, and the resulting heat may cause heat burns or ignite flammables. It also produces heat when reacted with acids.

Sodium hydroxide is corrosive to several metals, like aluminium which reacts with the alkali to produce flammable hydrogen gas on contact:

$$2\,Al + 6\,NaOH + \rightarrow 3\,H_2 + 2\,Na_3AlO_3$$

$$2\,Al + 2\,NaOH + 2\,H_2O \rightarrow 3\,H_2 + 2\,NaAlO_2$$

$$2\,Al + 2\,NaOH + 6\,H_2O \rightarrow 3\,H_2 + 2\,NaAl(OH)_4$$

Sodium hydroxide is also mildly corrosive to glass, which can cause damage to glazing or freezing of ground glass joints. Careful storage is needed.

Commercial Brands

- Red Devil Lye
- Lewis Lye

Potassium Hydroxide

Potassium hydroxide

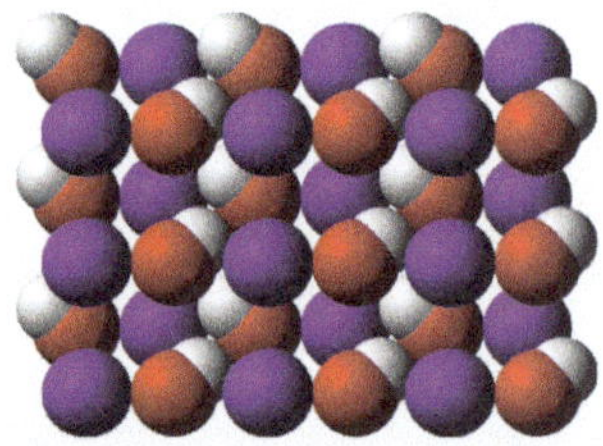

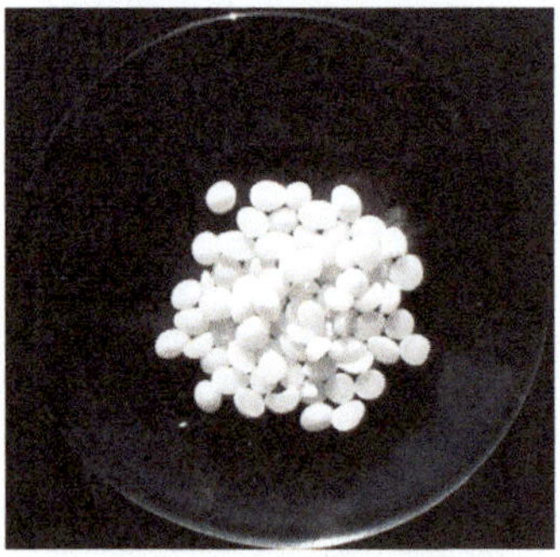

Potassium hydroxide is an inorganic compound with the formula KOH, and is commonly called caustic potash.

Along with sodium hydroxide (NaOH), this colorless solid is a prototypical strong base. It has many industrial and niche applications, most of which exploit its corrosive nature and its reactivity toward acids. An estimated 700,000 to 800,000 tonnes were produced in 2005. Approximately 100 times more NaOH than KOH is produced annually. KOH is noteworthy as the precursor to most soft and liquid soaps as well as numerous potassium-containing chemicals.

Properties and Structure

Potassium hydroxide can be found in pure form by reacting sodium hydroxide with impure potassium. It is usually sold as translucent pellets, which will become tacky in air because KOH is hygroscopic. Consequently, KOH typically contains varying amounts of water. Its dissolution in water is strongly exothermic. Concentrated aqueous solutions are sometimes called potassium lyes. Even at high temperatures, solid KOH does not dehydrate readily.

Potassium hydroxide solutions with concentrations of approximately 0.5 to 2.0% are irritating when coming into contact with the skin, while concentrations higher than 2% are corrosive.

Structure

At higher temperatures, solid KOH crystallizes in the NaCl crystal structure. The OH group is either rapidly or randomly disordered so that the OH−group is effectively a spherical anion of radius 1.53 Å (between Cl−and F−in size). At room temperature, the OH−groups are ordered and the environment about the K+centers is distorted, with K+—OH−distances ranging from 2.69 to 3.15 Å, depending on the orientation of the OH group. KOH forms a series of crystalline hydrates, namely the monohydrate KOH·H2O, the dihydrateKOH·2 H2O, and the tetrahydrate KOH·4 H2O.

Solubility and Desiccating Properties

Approximately 121 g of KOH will dissolve in 100 mL of water at room temperature compared with 100 g of NaOH in the same volume (on a molar basis, KOH is slightly less soluble than NaOH). Lower molecular weight alcohols such as methanol, ethanol, and propanols are also excellent solvents.

Because of its high affinity for water, KOH serves as a desiccant in the laboratory. It is often used to dry basic solvents, especially amines and pyridines: distillation of these basic liquids from a slurry of KOH yields the anhydrous reagent.

Thermal Stability

Like NaOH, KOH exhibits high thermal stability. The gaseous species is dimeric. Because of its high stability and relatively low melting point, it is often melt-cast as pellets or rods, forms that have low surface area and convenient handling properties.

Reactions

As a Base

KOH is highly basic, forming strongly alkaline solutions in water and other polar solvents. These solutions are capable of deprotonating many acids, even weak ones. In analytical chemistry, titrations using solutions of KOH are used to assay acids.

As a Nucleophile in Organic Chemistry

KOH, like NaOH, serves as a source of OH−, a highly nucleophilic anion that attacks polar bonds in both inorganic and organic materials. In perhaps its most well-known reaction, aqueous KOH saponifies esters:

$$KOH + RCO_2R' \rightarrow RCO_2K + R'OH$$

When R is a long chain, the product is called a potassium soap. This reaction is manifested by the "greasy" feel that KOH gives when touched — fats on the skin are rapidly converted to soap and glycerol.

Molten KOH is used to displace halides and other leaving groups. The reaction is especially useful for aromatic reagents to give the corresponding phenols.

Reactions with Inorganic Compounds

Complementary to its reactivity toward acids, KOH attacks oxides. Thus, SiO_2 is attacked by KOH to give soluble potassium silicates. KOH reacts with carbon dioxide to give bicarbonate:

$$KOH + CO_2 \rightarrow KHCO_3$$

Manufacture

Historically KOH was made by adding potassium carbonate (potash) to a strong solution of calcium hydroxide (slaked lime), leading to a metathesis reaction which caused calcium carbonate to precipitate, leaving potassium hydroxide in solution:

$$Ca(OH)_2 + K_2CO_3 \rightarrow CaCO_3 + 2\ KOH$$

Filtering off the precipitated calcium carbonate and boiling down the solution gives potassium hydroxide ("calcinated or caustic potash"). It was the most important method of producing potassium hydroxide until the late 19th century, when it was largely replaced by the current method of electrolysis of potassium chloride solutions. The method is analogous to the manufacture of sodium hydroxide:

$$2\ KCl + 2\ H_2O \rightarrow 2\ KOH + Cl_2 + H_2$$

Hydrogen gas forms as a by-product on the cathode; concurrently, an anodic oxidation of the chloride ion takes place, forming chlorine gas as a by-product. Separation of the anodic and cathodic spaces in the electrolysis cell is essential for this process.

USES KOH and NaOH can be used interchangeably for a number of applications, although in industry, NaOH is preferred because of its lower cost.

Precursor to Other Potassium Compounds

Many potassium salts are prepared by neutralization reactions involving KOH. The potassium salts of carbonate, cyanide, permanganate, phosphate, and various silicates are prepared by treating either the oxides or the acids with KOH. The high solubility of potassium phosphate is desirable in fertilizers.

Manufacture of Biodiesel

Although more expensive than using sodium hydroxide, KOH works well in the manufacture of biodiesel by transesterification of the triglycerides in vegetable oil. Glycerin from potassium hydroxide-processed biodiesel is useful as an inexpensive food supplement for livestock, once the toxic methanol is removed.

Manufacture of Soft Soaps

The saponification of fats with KOH is used to prepare the corresponding "potassium soaps," which are softer than the more common sodium hydroxide-derived soaps. Because of their softness and greater solubility, potassium soaps require less water to liquefy, and can thus contain more cleaning agent than liquefied sodium soaps.

As an Electrolyte

Potassium carbonate, formed from the hydroxide solution leaking from an alkaline battery

Aqueous potassium hydroxide is employed as the electrolyte in alkaline batteries based on nickel-cadmium, nickel-hydrogen, and manganese dioxide-zinc. Potassium hydroxide is preferred over sodium hydroxide because its solutions are more conductive. The Nickel Metal Hydride batteries in the Toyota Prius use a mixture of potassium hydroxide and sodium hydroxide. Nickel–iron batteries also use potassium hydroxide electrolyte.

Niche Applications

Like sodium hydroxide, potassium hydroxide attracts numerous specialized applications, virtually all of which rely on its properties as a strong chemical base with its consequent ability to degrade many materials. For example, in a process commonly referred to as "chemical cremation" or "resomation," potassium hydroxide hastens the decomposition of soft tissues, both animal and human, to leave behind only the bones and other hard tissues. Entomologists wishing to study the fine structure of insect anatomy may use a 10% aqueous solution of KOH to apply this process.

In chemical synthesis, the choice between the use of KOH and the use of sodium hydroxide, NaOH, is guided by the solubility of the resulting salt.

The corrosive properties of potassium hydroxide make it a useful ingredient in agents and preparations that clean and disinfect surfaces and materials that can themselves resist corrosion by KOH.

KOH is also used for semiconductor chip fabrication.

Potassium hydroxide is often the main active ingredient in chemical "cuticle removers" used in manicure treatments.

Because aggressive bases like KOH damage the cuticle of the hair shaft, potassium hydroxide is used to chemically assist the removal of hair from animal hides. The hides are soaked for several hours in a solution of KOH and water to prepare them for the unhairing stage of the tanning process. This same effect is also used to weaken human hair in preparation for shaving. Pre-shave products and some shave creams contain potassium hydroxide to force open the hair cuticle and to act as a hygroscopic agent to attract and force water into the hair shaft, causing further damage to the hair. In this weakened state, the hair is more easily cut by a razor blade.

Potassium hydroxide is used to identify some species of fungi. A 3–5% aqueous solution of KOH is applied to the flesh of a mushroom and the researcher notes whether or not the color of the flesh changes. Certain species of gilled mushrooms, boletes, polypores, and lichens are identifiable based on this color-change reaction.

Petroleum Refineries

Potassium hydroxide is also used in petroleum and natural gas refining for removal of organic aci

Sodium Methoxide

Sodium methoxide is a chemical compound with the formula CH_3ONa. This colorless solid, which is formed by the deprotonation of methanol, is a widely used reagent in industry and the laboratory. It is also a dangerously caustic base.

Sodium methoxide
$Na^{+}\ {}^{-}O{-}CH_3$

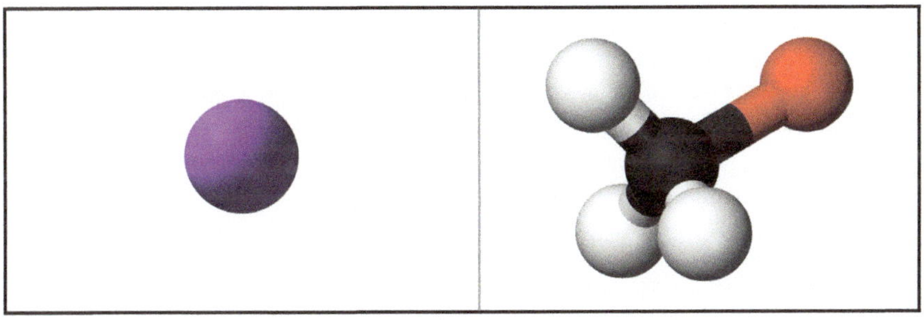

Preparation and Structure

Sodium methoxide is prepared by carefully treating methanol with sodium:

$$2\ Na + 2\ CH_3OH \rightarrow 2\ CH_3ONa + H_2$$

The reaction is so exothermic that ignition is possible. The resulting solution, which is colorless, is often used as a source of sodium methoxide, but the pure material can be isolated by evaporation followed by heating to remove residual methanol. The solid hydrolyzes in water to give methanol and sodium hydroxide; commercial-grade product can be contaminated with the hydroxide. The solid and especially solutions absorb carbon dioxide from the air, thus diminishing the effectiveness of the base.

In the solid form, sodium methoxide is polymeric, with a sheet-like arrays of Na^{+} centers, each bonded to four oxygen centers.

The structure, and hence its basicity, of sodium methoxide in solution depends on the solvent. It is a significantly stronger base in DMSO where it is more fully ionized and free of hydrogen bonding.

Applications

Organic Synthesis

Sodium methoxide is a routinely used base in organic chemistry, applicable to the synthesis of numerous compounds, ranging from pharmaceuticals to agrichemicals. As a base, it is employed in dehydrohalogenations and various condensations. It is also a nucleophile for the production of methyl ethers.

Industrial Applications

Sodium methoxide is used as an initiator of anionic addition polymerization with ethylene oxide, forming a polyether with high molecular weight. Biodiesel is prepared from vegetable oils and animal fats, that is, fatty acid triglycerides, by transesterification with methanol to give fatty acid methyl esters (FAMEs). This transformation is catalyzed by sodium methoxide.

Safety

Sodium methoxide is highly caustic, and the hydrolysis gives methanol, which is toxic and volatile.

NFPA 704

The ratings for this substance vary widely.

Rating	 4 3 3	 4 3 2	 3 3 2	 2 3 2	 2 2 2
Source	State of Connecticut	DuPont	Pharmco AAPR	ScienceLab(Both ratings on same sheet)	

References

- Zumdahl, Steven S. (2009). Chemical Principles 6th Ed. Houghton Mifflin Company. p. A23. ISBN 0-618-94690-X.
- Roach, Mary (2004). Stiff: The Curious Lives of Human Cadavers, New York: W.W. Norton & Company. ISBN 0-393-32482-6.
- Lide, D. R., ed. (2005). CRC Handbook of Chemistry and Physics (86th ed.). Boca Raton (FL): CRC Press. p. 4-80. ISBN 0-8493-0486-5.
- Zumdahl, Steven S. (2009). Chemical Principles 6th Ed. Houghton Mifflin Company. p. A22. ISBN 0-618-94690-X.
- Elix, J.A.; Stocker-Wörgötter, Elfie (2008). "Chapter 7: Biochemistry and secondary metabolites". In Nash III, Thomas H. Lichen Biology (2nd ed.). New York: Cambridge University Press. pp. 118–119. ISBN 978-0-521-69216-8.
- Brian Oram, PG (2014). "Drinking Water Issues Corrosive Water (Lead, Copper, Aluminum, Zinc and More)". Retrieved June 23, 2016.
- Seidell, Atherton; Linke, William F. (1952). Solubilities of Inorganic and Organic Compounds. Van Nostrand. Retrieved 2014-05-29.

- Ayres, Chris (27 February 2010) Clean green finish that sends a loved one down the drain Times Online. Retrieved 2013-02-20.
- "aluminium_water_hydrogen.pdf (application/pdf Object)" (PDF). www1.eere.energy.gov. 2008. Retrieved January 15, 2013.
- Andrew Hough (18 October 2012). "British engineers produce amazing 'petrol from air' technology". The Telegraph. Retrieved 21 October 2012.
- "Olives: Safe Methods for Home Pickling (application/pdf Object)" (PDF). ucanr.org. 2010. Retrieved January 22, 2012.
- Green, Margaret (January 1952). "A RAPID METHOD FOR CLEARING AND STAINING SPECIMENS FOR THE DEMONSTRATION OF BONE" (PDF). The Ohio Journal of Science. 52 (1): 31–33. Retrieved 20 November 2012.

6

Essential Aspects of Biodiesel Production

The essential aspects elucidated in this section are cellulosic ethanol commercialization, energy returned on energy invested, Fischer-Speier esterification and glycerol. The method of converting cellulose-containing organic matter into cellulosic ethanol for using it as a biofuel is known as cellulosic ethanol commercialization. The chapter closely examines the aspects of biodiesel to provide the reader an extensive understanding of the subject.

Cellulosic Ethanol Commercialization

Cellulosic ethanol commercialization is the process of building an industry out of methods of turning cellulose-containing organic matter into cellulosic ethanol for use as a biofuel. Companies, such as Iogen, POET, DuPont, and Abengoa, are building refineries that can process biomass and turn it into bioethanol. Companies, such as Diversa, Novozymes, and Dyadic, are producing enzymes that could enable a cellulosic ethanol future. The shift from food crop feedstocks to waste residues and native grasses offers significant opportunities for a range of players, from farmers to biotechnology firms, and from project developers to investors.

As of 2013, the first commercial-scale plants to produce cellulosic biofuels have begun operating. Multiple pathways for the conversion of different biofuel feedstocks are being used. In the next few years, the cost data of these technologies operating at commercial scale, and their relative performance, will become available. Lessons learnt will lower the costs of the industrial processes involved.

Commercialization by Country

Australia

Ethtec is building a pilot plant in Harwood, New South Wales, which uses wood residues as a feedstock.

Cellulosic Ethanol Production

Cellulosic ethanol can be produced from a diverse array of feedstocks, such as wood pulp from trees or any plant matter. Instead of taking the grain from wheat and grinding that down to get starch and gluten, then taking the starch, cellulosic ethanol production involves the use of the whole crop. This approach should increase yields and reduce the carbon footprint because the amount of energy-intensive fertilisers and fungicides will remain the same, for a higher output of usable material.

Brazil

GranBio (formerly known as GraalBio) is building a facility projected to produce 82 million litres of cellulosic ethanol per year.

Canada

In Canada, Iogen Corp. is a developer of cellulosic ethanol process technology. Iogen has developed a proprietary process and operates a demonstration-scale plant in Ontario. The facility has been designed and engineered to process 40 tons of wheat straw per day into ethanol using enzymes made in an adjacent enzyme manufacturing facility. In 2004, Iogen began delivering its first shipments of cellulosic ethanol into the marketplace. In the near term, the company intends to commercialize its cellulose ethanol process by licensing its technology broadly through turnkey plant construction partnerships. The company is currently evaluating sites in the United States and Canada for its first commercial-scale plant.

Lignol Innovations has a pilot plant, which uses wood as a feedstock, in Vancouver.

In March 2009, KL Energy Corporation of South Dakota and Prairie Green Renewable Energy of Alberta announced their intention to develop a cellulosic ethanol plant near Hudson Bay, Saskatchewan. The Northeast Saskatchewan Renewable Energy Facility will use KL Energy's modern design and engineering to produce ethanol from wood waste.

China

Cellulosic ethanol production currently exists at "pilot" and "commercial demonstration" scale, including a plant in China engineered by SunOpta Inc. and owned and operated by China Resources Alcohol Corporation that is currently producing cellulosic ethanol from corn stover (stalks and leaves) on a continuous, 24-hour-per-day basis.

Denmark

Inbicon's bioethanol plant in Kalundborg, with the capacity to produce 5.4 million liters (1.4 million gallons) annually, was opened in 2009. Believed to be the world's largest cellulosic ethanol plant as of early 2011, the facility runs on about 30,000 metric tons (33,000 tons) of straw per year and the plant employs about 30 people. The plant also produces 13,000 metric tons of lignin pellets per year, used as fuel at combined-heat-and-power plants, and 11,100 metric tons of C5 molasses which is currently used for biomethane production via anaerobic digestion, and has been tested as a high carbohydrate animal feed supplement and potential bio-based feedstock for production of numerous commodity chemicals including diols, glycols, organic acids, and biopolymer precursors and intermediates.

Since October 2010, an E5 blend of 95% gasoline and 5% cellulosic ethanol blend has been available at 100 filling stations across Denmark. Distributed by Statoil, the Bio95 2G mixture uses ethanol derived from wheat straw collected on Danish fields after harvest and produced by Inbicon (a div. of DONG Energy), using enzyme technology from Novozymes.

Commercial or experimental Cellulosic Ethanol Plants in Denmark (Operational or under construction)				
Company	**Location**	**Feedstock**	**Yearly amount**	**Operational**
Biogasol	Bornholm	Wheat straw	5 mill litre	2012
Ensted-værket	Aabenraa	Wheat straw	? mill litre	2013
Inbicon owned by Dong Energy	Kalundborg, Zealand	Wheat straw	5.4 mill litre	2009
Maabjerg Energy Concept owned by Dong Energy	Maabjerg	Wheat straw	50-70 mill litre	Canceled 2016

Germany

The biofuel company Butalco has recently signed a research and development contract with Hohenheim University. The Institute of Fermentation Technology within the Department of Food Science and Biotechnology at Hohenheim University has been concerned with questions on the production of bioethanol for almost 30 years. The focus in recent years has been on the improvement of the material, energy and life cycle assessment of the production of ethanol. Special interest to BUTALCO is the use of the newly built pilot plant, which is equipped with a safety class 1 approved fermentation room with 4 x 1.5 m^3 fermenters. The concept of the plant allows both starch and lignocellulosic based raw materials to be processed. The collaboration will allow BUTALCO to optimise its C5 sugar fermenting and butanol producing yeast strains on a technical scale and produce first amounts of bioethanol from lignocellulose. The whole process of the production of biofuel from the choice of cellulosic biomass feedstock to the conversion into sugars and fermentation through to the purification will be optimised under industrial conditions.

India

Cellulosic ethanol production currently exists at "pilot" scale, with efforts being made on utilization of waste lignocellulosic biomass for ethanol production. Pilot scale studies for utilization of pine needles and Lantana weed undertaken at Cellulose and Paper Division, Forest Research Institute, Dehradun, India.

Italy

Italy-based Mossi & Ghisolfi Group broke ground for its 13 million US gallons (49,000 m^3) per year cellulosic ethanol facility in Crescentino in northwestern Italy on April 12, 2011. The project will be the largest cellulosic ethanol project in the world, 10 times larger than any of the currently operating demonstration-scale facilities. The plant is "expected to become operational in 2012 and will use a variety of locally sourced feedstocks, beginning with wheat straw and Arundo donax, a perennial giant cane".

Japan

Nippon Oil Corporation and other Japanese manufacturers including Toyota Motor Corporation plan to set up a research body to develop cellulose-derived biofuels. The consortium plans to produce 250,000 kilolitres (1.6 million barrels) per year of bioethanol by March 2014, and produce bioethanol at 40 yen ($0.437) per litre (about $70 a barrel) by 2015.

In March 2009, Honda Motor announced an agreement for the construction of a new cellulosic ethanol research facility in Japan. The new Kazusa-branch facility of the Honda Fundamental Technology Research Center will be built within the Kazusa Akademia Park, in Kisarazu, Chiba. Construction is scheduled to begin in April 2009, with the aim to begin operations in November 2009.

Norway

In October 2010, Norway-based cellulosic ethanol technology developer Weyland commenced production at its 200,000 liter (approximately 53,000 gallon) pilot-scale facility in Bergen, Norway. The plant will demonstrate the company's acid hydrolysis production process, paving the way for a commercial-scale project. The company also plans to market its technology worldwide.

Russia

A commercial factory converting wood (50% softwood + 50% hardwood) into Ethanol is in operation in Northern Russia, the city of Kirov, since 1972 and is still profitable. As side products the company, Kirov Biochemical Works, is offering dry fodder yeast (20 tons/month) and Lignin. To install equipment for drying and burning Lignin, both fresh and accumulated in the landfill, for steam and electricity, a bank loan of $200 million was recently secured.

Spain

Abengoa continues to invest heavily in the necessary technology for bringing cellulosic ethanol to market. Utilizing process and pre-treatment technology from SunOpta Inc., Abengoa is building a 5 million US gallons (19,000 m^3) cellulosic ethanol facility in Spain and have recently entered into a strategic research and development agreement with Dyadic International, Inc. (AMEX: DIL), to create new and better enzyme mixtures which may be used to improve both the efficiencies and cost structure of producing cellulosic ethanol.

Sweden

SEKAB has developed an industrial process for production of ethanol from biomass feed-stocks, including wood chips and sugar cane bagasse. The development work is being carried out at an advanced pilot plant in Örnsköldsvik, and has sparked international interest. The technology will be gradually scaled up to commercial production in a new breed of bio-refineries from 2013 to 2015.

United States

The US government actively supports the development and commercialization of cellulosic ethanol through a variety of mechanisms. In the first decade of the 21st century, a lot of companies announced plans to build commercial cellulosic ethanol plants, but most of those plans eventually fell apart, and many of the small companies went bankrupt. By today (2012), there are many demonstration plants throughout the country, and handful of commercial-scale plants which are in operation or close to it. With the market for cellulosic ethanol in the United States projected to continue growing in the coming years, the outlook for this industry is good.

Government Support

The US Federal government is actively promoting the development of ethanol from cellulosic feedstocks as an alternative to conventional petroleum transportation fuels. For example, programs sponsored by U.S. Department of Energy (DOE) include research to develop better cellulose hydrolysis enzymes and ethanol-fermenting organisms, to engineering studies of potential processes, to co-funding initial ethanol from cellulosic biomass demonstration and production facilities. This research is conducted by various national laboratories,including the National Renewable Energy Laboratory (NREL), Oak Ridge National Laboratory (ORNL) and Idaho National Laboratory (INL), as well as by universities and private industry. Engineering and construction companies and operating companies are generally conducting the engineering work.

In May 2008, Congress passed a new farm bill that will accelerate the commercialization of advanced biofuels, including cellulosic ethanol. The *Food, Conservation, and Energy Act of 2008* provides for grants covering up to 30% of the cost of developing and building demonstration-scale biorefineries for producing "advanced biofuels," which essentially includes all fuels that are not produced from corn kernel starch. It also allows for loan guarantees of up to $250 million for building commercial-scale biorefineries to produce advanced biofuels.

Using a newly developed tool known as the "Biofuels Deployment Model", Sandia researchers have determined that 21 billion US gallons (79,000,000 m^3) of cellulosic ethanol could be produced per year by 2022 without displacing current crops. The Renewable Fuels Standard, part of the 2007 Energy Independence and Security Act, calls for an increase in biofuels production to 36 billion US gallons (140,000,000 m^3) a year by 2022.

In January 2011, the USDA approved $405 million in loan guarantees through the 2008 Farm Bill to support the commercialization of cellulosic ethanol at three facilities owned by Coskata, Enerkem and INEOS New Planet BioEnergy. The projects represent a combined 73 million US gallons (280,000 m^3) per year production capacity and will begin producing cellulosic ethanol in 2012. The USDA also released a list of advanced biofuel producers who will receive payments to expand the production of advanced biofuels. In July 2011, the US Department of Energy gave in $105 million in loan guarantees to POET for a commercial-scale plant to be built Emmetsburg, Iowa.

Commercial Development

The cellulosic ethanol industry in the United States developed some new commercial-scale plants in 2008. Plants totaling 12 million liters (3.17 million gal) per year were operational, and an additional 80 million liters (21.13 million gal.) per year of capacity - in 26 new plants - was under construction. (For comparison the estimated US petroleum consumption for all uses was about 816 million gal/day in 2008.)

Cellulosic Ethanol Plants in the U.S. (Operational or under construction)					
Company	**Location**	**Feedstock**	**Capacity (million gal/year)**	**Began Production**	**Type**
Abengoa Bioenergy	Hugoton, KS	Wheat straw	25 - 30	est. late 2013	Commercial
American Process, Inc	Alpena, MI	Wood chips	1.0		Commercial
BlueFire Ethanol	Irvine, CA	Multiple sources	3.9		Commercial
BlueFire Ethanol	Fulton, MS	Multiple sources	19		Commercial
Coskata, Inc.	Madison, Pennsylvania	Multiple sources	0.04	October 2009	Semi-commercial
DuPont	Nevada, IA	Corn stover	30	est. 2014	Commercial
Fulcrum BioEnergy	Reno, NV	Municipal solid waste	10	est. end of 2013	Commercial

Cellulosic Ethanol Plants in the U.S. (Operational or under construction)					
Company	**Location**	**Feedstock**	**Capacity (million gal/year)**	**Began Production**	**Type**
Gulf Coast Energy	Livingston, AL	Wood waste	0.3	before 2008	Demonstration
KL Energy Corp.	Upton, WY	Wood waste			
Mascoma	Kinross, MI	Wood waste	20		Commercial
POET LLC	Emmetsburg, IA	Corn stover	20 - 25	Sept. 2014	Commercial
POET LLC	Scotland, SD	Corn stover	0.03	2008	Pilot

Environmental Issues

Cellulosic ethanol and grain-based ethanol are, in fact, the same product, but many scientists believe cellulosic ethanol production has distinct environmental advantages over grain-based ethanol production. On a life-cycle basis, ethanol produced from agricultural residues or dedicated cellulosic crops has significantly lower greenhouse gas emissions and a higher sustainability rating than ethanol produced from grain.

According to US Department of Energy studies conducted by the Argonne National Laboratory of the University of Chicago, cellulosic ethanol reduces greenhouse gas emissions (GHG) by 85% over reformulated gasoline. By contrast, starch ethanol (e.g., from corn), which usually uses natural gas to provide energy for the process, reduces greenhouse gas emissions by 18% to 29% over gasoline.

Critics such as Cornell University professor of ecology and agriculture David Pimentel and University of California at Berkeley engineer Tad Patzek question the likelihood of environmental, energy, or economic benefits from cellulosic ethanol technology from non-waste.

Energy Returned on Energy Invested

In physics, energy economics, and ecological energetics, energy returned on energy invested (EROEI or ERoEI); or energy return on investment (EROI), is the ratio of the amount of usable energy delivered from a particular energy resource to the amount of energy used to obtain that energy resource. It is a distinct measure from energy efficiency as it does not measure the primary energy inputs to the system, only usable energy.

A fuel or energy must have an EROEI ratio of at least 3:1 to be considered viable as a prominent fuel or energy source.

$$EROEI = \frac{\text{Energy Delivered}}{\text{Energy Required to Deliver that Energy}}$$

When the EROEI of a resource is less than or equal to one, that energy source becomes a net "energy sink", and can no longer be used as a source of energy, but depending on the system might be useful for energy storage (for example a battery). A related measure Energy Store On Energy Invested (ESOEI) is used to analyse storage systems.

Non-manmade Energy Inputs

The natural or primary energy sources are not usually included in the calculation of energy invested, only the human-applied sources. For example, in the case of biofuels the solar insolation driving photosynthesis is not included, and the energy used in the stellar synthesis of fissile elements is not included for nuclear fission. The energy returned includes usable energy and not wastes such as heat, although depending on source and application, waste heat is used in district heating and water desalination, these cogeneration plants however are rare, globally, and thus usually excluded in EROEI analysis of energy sources.

Relationship to Net Energy Gain

EROEI and *Net energy (gain)* measure the same quality of an energy source or sink in numerically different ways. Net energy describes the amounts, while EROEI measures the ratio or efficiency of the process. They are related simply by

$$\text{GrossEnergyYield} \div \text{EnergyExpended} = EROEI$$

or

$$(\text{NetEnergy} \div \text{EnergyExpended}) + 1 = EROEI$$

For example, given a process with an EROEI of 5, expending 1 unit of energy yields a net energy gain of 4 units. The break-even point happens with an EROEI of 1 or a net energy gain of 0.

Low Carbon Power

A 2015 review in Renewable and Sustainable Energy Reviews assessed the energy payback time and EROI of solar photovoltaics. In this study, which uses an insolation of $1700/kWh/m^2/yr$ and a system lifetime of 30 years, mean harmonized EROIs between 8.7 and 34.2 were found. Mean harmonized energy payback time varied from 1.0 to 4.1 years. A review Pickard reports EROEI estimates for monocrystalline silicon photovoltaics by four groups in the range of 2.2 to 8.8. It should be noted that EROEI is a function of the solar irradiation level and the type of PV technology; it also depends on methodology and system boundary assumptions. Raugei, Fullana-i-Palmer and Fthenakis show EROEI in the range of 5.9 to 11.8 and 19 to 39 for the major commercial PV types in South European installations. The low range assumes that primary energy and electricity are of the same quality, whereas the high range (19-39) is calculated by converting the electricity output of PV to primary energy as recommended by the IEA PVPS Task 12 LCA Methodology Guidelines. Furthermore, Fthenakis determined the EROEI to be as high as 60 for the least energy consuming thin-film PV technology installations in the U.S. Southwest.

The EROI of wind turbines depends on invested energy in the turbine, produced energy and life span of a turbine. In the scientific literature EROIs normally vary between 20 and 50.

Economic Influence of EROEI

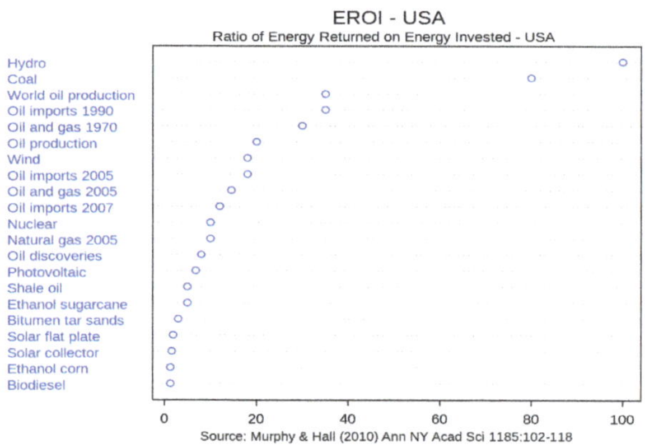

EROI — Ratio of Energy Returned on Energy Invested. Assuming, amongst other things, "Nuclear" depended upon diffusion enrichment which was still used in the US at the time of publication - Murphy & Hall 2010. In 2013 however, the last uranium diffusion plant in the world, at Paducah, closed down. Only newer more efficient technology is now used.

EROI (for US)	Fuel
1.3	Biodiesel
3.0	Bitumen tar sands
80.0	Coal
1.3	Ethanol corn
5.0	Ethanol sugarcane
100.0	Hydro
35.0	Oil imports 1990

EROI (for US)	Fuel
18.0	Oil imports 2005
12.0	Oil imports 2007
8.0	Oil discoveries
20.0	Oil production
10.0	Natural gas 2005
75	Nuclear (Centrifugal enrichment)
10.0	Nuclear (with diffusion enrichment- Obsolete)
2000 (estimate)	Dual Fluid Molten salt - Molten lead Nuclear
30.0	Oil and gas 1970
14.5	Oil and gas 2005
6.8	Photovoltaic
5.0	Shale oil
1.6	Solar collector
1.9	Solar flat plate
18.0	Wind
9.5	Geothermal (without hot water heating)
32.4	Geothermal (with hot water heating)

High per-capita energy use has been considered desirable as it is associated with a high standard of living based on energy-intensive machines. A society will generally exploit the highest available EROEI energy sources first, as these provide the most energy for the least effort. This is an example of David Ricardo's best-first principle. Then progressively lower quality ores or energy resources are used as the higher-quality ones are either exhausted or in use, for example, wind turbines positioned in the windiest areas.

In regard to fossil fuels, when oil was originally discovered, it took on average one barrel of oil to find, extract, and process about 100 barrels of oil. The ratio, for discovery of fossil fuels in the United States, has declined steadily over the last century from about 1000:1 in 1919 to only 5:1 in the 2010s.

Although many qualities of an energy source matter (for example oil is energy-dense and transportable, while wind is variable), when the EROEI of the main sources of energy for an economy fall that energy becomes more difficult to obtain and its relative price increases. Therefore, the EROEI gains importance when comparing energy alternatives. Since expenditure of energy to obtain energy requires productive effort, as the EROEI falls an increasing proportion of the economy has to be devoted to obtaining the same amount of net energy.

Since the invention of agriculture, humans have increasingly used exogenous sources of energy to multiply human muscle-power. Some historians have attributed this largely to more easily exploited (i.e. higher EROEI) energy sources, which is related to the concept of energy slaves. Thomas Homer-Dixon argues that a falling EROEI in the Later Roman Empire was one of the reasons for the collapse of the Western Empire in the fifth century CE. In "The Upside of Down" he suggests that EROEI analysis provides a basis for the analysis of the rise and fall of civilisations. Looking at the maximum extent of the Roman Empire, (60 million) and its technological base the agrarian base of Rome was about 1:12 per hectare for wheat and 1:27 for alfalfa (giving a 1:2.7 production for oxen). One can then use this to calculate the population of the Roman Empire required at its height, on the basis of about 2,500–3,000 calories per day per person. It comes out roughly equal to the area of food production at its height. But ecological damage (deforestation, soil fertility loss particularly in southern Spain, southern Italy, Sicily and especially north Africa) saw a collapse in the system beginning in the 2nd century, as EROEI began to fall. It bottomed in 1084 when Rome's population, which had peaked under Trajan at 1.5 million, was only 15,000. Evidence also fits the cycle of Mayan and Cambodian collapse too. Joseph Tainter suggests that diminishing returns of the EROEI is a chief cause of the collapse of complex societies, this has been suggested as caused by peak wood in early societies. Falling EROEI due to depletion of high quality fossil fuel resources also poses a difficult challenge for industrial economies.

Tim Garrett links EROEI and inflation directly, based on a thermodynamic analysis of historical world energy consumption (Watts) and accumulated global wealth (US dollars). This economic growth model indicates that global EROEI is the inverse of global inflation over a given time interval. Because the model aggregates supply chains globally, local EROEI is outside its scope.

Oil Sands

Because much of the energy required for producing oil from oil sands (bitumen) comes from low value fractions separated out by the upgrading process, there are two ways to calculate EROEI, the higher value given by considering only the external energy inputs and the lower by considering all energy inputs, including self generated. Oil sands#Input energy "utilized detailed energy production and consumption data reported by oil sands producers from 1970 to 2010 to examine trends in historical energy returns from oil sands extraction. " They argued that by 2010, NERs (net energy returns) from oil sands mining and in situ operations had become significantly more energy efficient since 1970 although the NER remained significantly less efficient than conventional oil production. NERs from the oil sands, grew from "1.0 GJ/GJ in 1970 (entirely from the Suncor mining operation) to 2.95 GJ/GJ in 1990 and then to 5.23 GJ/GJ in 2010."

Criticism of EROEI

Measuring the EROEI of a single physical process is unambiguous, but there is no agreed-upon standard on which activities should be included in measuring the EROEI of an economic process. In addition, the form of energy of the input can be completely different from the output. For example, energy in the form of coal could be used in the production of ethanol. This might have an EROEI of less than one, but could still be desirable due to the benefits of liquid fuels (assuming the latters are not used in the processes of extraction and transformation).

How deep should the probing in the supply chain of the tools being used to generate energy go? For example, if steel is being used to drill for oil or construct a nuclear power plant, should the energy input of the steel be taken into account, should the energy input into building the factory being used to construct the steel be taken into account and amortized? Should the energy input of the roads which are used to ferry the goods be taken into account? What about the energy used to cook the steelworker's breakfasts? These are complex questions evading simple answers. A full accounting would require considerations of opportunity costs and comparing total energy expenditures in the presence and absence of this economic activity.

However, when comparing two energy sources a standard practice for the supply chain energy input can be adopted. For example, consider the steel, but don't consider the energy invested in factories deeper than the first level in the supply chain.

Energy return on energy invested does not take into account the factor of time. Energy invested in creating a solar panel may have consumed energy from a high power source like coal, but the return happens very slowly, i.e. over many years. If energy is increasing in relative value this should favour delayed returns. Some believe this means the EROEI measure should be refined further.

Conventional economic analysis has no formal accounting rules for the consideration of waste products that are created in the production of the ultimate output. For example, differing economic and energy values placed on the waste products generated in the production of ethanol makes the calculation of this fuel's true EROEI extremely difficult.

EROEI is only one consideration and may not be the most important one in energy policy. Energy independence (reducing international competition for limited natural resources), decrease of greenhouse gas emissions (including carbon dioxide and others), and affordability could be more important, particularly when considering secondary energy sources. While a nation's primary energy source is not sustainable unless it has a use rate less than or equal to its replacement rate, the same is not true for secondary energy supplies. Some of the energy surplus from the primary energy source can be used to create the fuel for secondary energy sources, such as for transportation.

Richards and Watt propose an *Energy Yield Ratio* for photovoltaic systems as an alternative to EROEI (which they refer to as *Energy Return Factor*). The difference is that it uses the design lifetime of the system, which is known in advance, rather than the actual lifetime. This also means that it can be adapted to multi-component systems where the components have different lifetimes.

Another issue with EROI that many studies attempt to tackle is that the energy returned can be in different forms, and these forms can have different utility. For example, electricity can be converted more efficiently than thermal energy into motion, due to electricity's lower entropy.

Additional EROEI Calculations

There are three prominent expanded EROEI calculations, they are point of use, extended and societal. Point of Use EROEI expands the calculation to include the cost of refining and transporting the fuel during the refining process. Since this expands the bounds of the calculation to include more production process EROEI will decrease. Extended EROEI includes point of use expansions as well as including the cost of creating the infrastructure needed for transportation of the energy or fuel once refined. Societal EROI is a sum of all the EROEIs of all the fuels used in a society or nation. A societal EROI has never been calculated and researchers believe it may currently be impossible to know all variables necessary to complete the calculation, but attempted estimates have been made for some nations. Calculations done by summing all of the EROEIs for domestically produced and imported fuels and comparing the result to the Human Development Index (HDI), a tool often used to understand well-being in a society. According to this calculation, the amount of energy a society has available to them increases the quality of life for the people living in that country and countries with less energy available also have a harder time satisfying citizens' basic needs. This is to say that societal EROI and overall quality of life are very closely linked.

ESOEI

ESOEI is used when EROEI is below 1.

Storage Technology	ES-OEI
Lithium ion battery	32
NaS battery	20
Lead acid battery	5
Vanadium redox battery	10
Zinc bromide battery	9
CAES	792
Pumped hydroelectric storage	704

EROEi Under Rapid Growth

A related recent concern is energy cannibalism where energy technologies can have a limited growth rate if climate neutrality is demanded. Many energy technologies are capable of replacing significant volumes of fossil fuels and concomitant green house gas emissions. Unfortunately, neither the enormous scale of the current fossil fuel energy system nor the necessary growth rate of these technologies is well understood within the limits imposed by the net energy produced for a growing industry. This technical limitation is known as energy cannibalism and refers to an effect where rapid growth of an entire energy producing or energy efficiency industry creates a need for energy that uses (or cannibalizes) the energy of existing power plants or production plants.

The solar breeder overcomes some of these problems. A solar breeder is a photovoltaic panel manufacturing plant which can be made energy-independent by using energy derived from its own roof using its own panels. Such a plant becomes not only energy self-sufficient but a major supplier of new energy, hence the name solar breeder. Research on the concept was conducted by Centre for Photovoltaic Engineering, University of New South Wales, Australia. The reported investigation establishes certain mathematical relationships for the solar breeder which clearly indicate that a vast amount of net energy is available from such a plant for the indefinite future. The solar module processing plant at Frederick, Maryland was originally planned as such a solar breeder. In 2009 the Sahara Solar Breeder Project was proposed by the *Science Council of Japan* as a cooperation between Japan and Algeria with the highly ambitious goal of creating hundreds of GW of capacity within 30 years. Theoretically breeders of any kind can be developed. In practice, nuclear breeder reactors are the only large scale breeders that have been constructed as of 2014, with the 600 MWe BN-600 and 800 MWe BN-800 reactor, the two largest in operation.

Fischer–Speier Esterification

Fischer–Speier esterification	
Named after	Hermann Emil Fischer Arthur Speier
Reaction type	Coupling reaction
Identifiers	
Organic Chemistry Portal	fischer-esterification
RSC ontology ID	RXNO:0000167

Fischer esterification or Fischer–Speier esterification is a special type of esterification by refluxing a carboxylic acid and an alcohol in the presence of an acid catalyst. The reaction was first described by Emil Fischer and Arthur Speier in 1895. Most carboxylic acids are suitable for the reaction, but the alcohol should generally be a primary or secondary alkyl. Tertiary alcohols are prone to elimination, and phenols are usually too unreactive to give useful yields. Commonly used catalysts for a Fischer esterification include sulfuric acid, tosylic acid, and Lewis acids such as scandium(III) triflate. For more valuable or sensitive substrates (for example, biomaterials), dicyclohexylcarbodiimide is often used. The reaction is often carried out without a solvent (particularly when a large reagent excess of alcohol is used) or in a non-polar solvent (e.g. toluene) to facilitate the Dean-Stark method. Typical reaction times vary from 1–10 hours at temperatures of 60-110 °C.

The synthesis of methyl benzoate by Fischer–Speier esterification

Direct acylations of alcohols with carboxylic acids is preferred over acylations with anhydrides (poor atom economy) or acid chlorides (moisture sensitive). The main disadvantage of direct acylation is the unfavorable chemical equilibrium that must be remedied (e.g. by a large excess of one of the reagents), or by the removal of water (e.g. by using Dean-Stark distillation, anhydrous salts, molecular sieves, or by using a stoichiometric quantity of acid catalyst).

Overview

Fischer esterification is an example of nucleophilic acyl substitution based on the electrophilicity of the carbonyl carbon and the nucleophilicity of an alcohol. However, carboxylic acids tend to be less reactive than esters as electrophiles. Additionally, in dilute neutral solutions they tend to be deprotonated anions (and thus unreactive as electrophiles). Though very kinetically slow without any catalysts (most esters are metastable), pure esters will tend to spontaneously hydrolyse in the presence of water, so when carried out "unaided", high yields for this reaction is quite unfavourable.

Several steps can be taken to turn this unfavourable reaction into a favourable one.

The reaction mechanism for this reaction has several steps:

1. Proton transfer from acid catalyst to carbonyl oxygen increases electrophilicity of carbonyl carbon.
2. The carbonyl carbon is then attacked by the nucleophilic oxygen atom of the alcohol
3. Proton transfer from the oxonium ion to a second molecule of the alcohol gives an activated complex

4. Protonation of one of the hydroxyl groups of the activated complex gives a new oxonium ion.
5. Loss of water from this oxonium ion and subsequent deprotonation gives the ester.

A generic mechanism for an acid Fischer esterification is shown below.

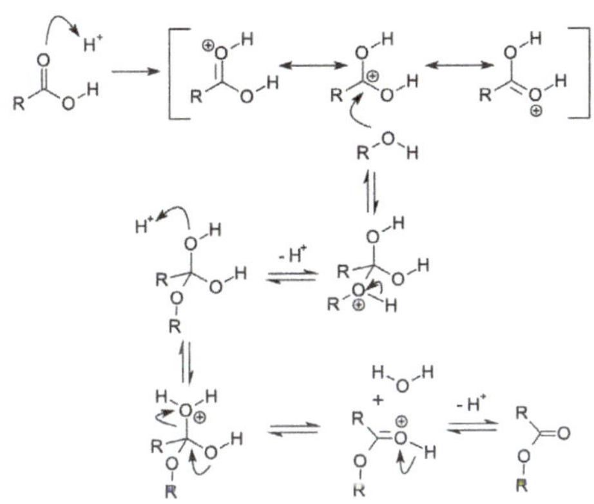

Advantages and Disadvantages

Advantages

The primary advantages of Fischer esterification compared to other esterification processes are based on its relative simplicity. Straightforward acidic conditions can be used if acid-sensitive functional groups are not an issue; sulfuric acid can be used; softer acids can be used with a tradeoff of longer reaction times. Because the reagents used are "direct," there is less environmental impact in terms of waste products and harmfulness of the reagents. Alkyl halides are potential greenhouse gases or ozone depletors and possible ecological poisons. Acid chlorides evolve hydrogen chloride gas upon contact with atmospheric moisture, are corrosive, react vigorously with water and other nucleophiles (sometimes dangerously); they are easily quenched by other nucleophiles besides the desired alcohol; their most common synthesis routes involve the evolution of toxic carbon monoxide or sulfur dioxide gases (depending on the synthesis process used).

Acid anhydrides are more reactive than esters because the leaving group is a carboxylate anion—a better leaving group than an alkoxide anion because their negative charge is more delocalised. However, such routes generally result in poor atom economy. For example, in reacting ethanol with acetic anhydride, ethyl acetate forms and acetic acid is kicked out as a leaving group, which is considerably less reactive than an acid anhydride and will be left as a byproduct (in a wasteful 1:1 ratio with the ester product) if product is collected immediately. If conditions are acidic enough, the acetic acid can be further reacted via the Fischer esterification pathway, but at a much slower pace.

However, in many carefully designed syntheses, reagents can be designed such that acid anhydrides are generated in situ and carboxylic acid byproducts are reactivated, and Fischer esterification routes are not necessarily mutually exclusive with acetic anhydride routes. (Examples of this include the common undergraduate organic lab experiment involving the acetylation of salicylic acid to yield aspirin.)

Fischer esterification is primarily a thermodynamically-controlled process: because of its slowness, the most stable ester tends to be the major product. This can be a desirable trait if there are multiple reaction sites and side product esters to be avoided. In contrast, rapid reactions involving acid anhydrides or acid chlorides are often kinetically-controlled.

Disadvantages

The primary disadvantages of Fischer esterification routes are its thermodynamic reversibility and relatively slow reaction rates—often on the scale of several hours to years, depending on the reaction conditions. Workarounds to this can be inconvenient if there are other functional groups sensitive to strong acid, in which case other catalytic acids may be chosen. If the product ester has a lower boiling point than either water or the reagents, the product may be distilled rather than water; this is common as esters with no protic functional groups tend to have lower boiling points than their protic parent reagents. Purification and extraction are easier if the ester product can be distilled away from the reagents and byproducts, but reaction rate can be slowed because overall reaction temperature can be limited in this scenario. A more inconvenient scenario is if the reagents have a lower boiling point than either the ester product or water, in which case the reaction mixture must be capped and refluxed and a large excess of starting material added. In this case anhydrous salts, such as Copper(II) sulfate or Potassium pyrosulfate, can also be added to sequester the water by forming hydrates, shifting the equilibrium toward ester products. These hydrated salts are then decanted prior to the final workup.

Examples in Alcoholic Beverages

The natural esterification that takes place in wines and other alcoholic beverages during the aging process is an example of acid-catalysed esterification. Over time, the acidity of the acetic acid and tannins in an aging wine will catalytically protonate other organic acids (including acetic acid itself), encouraging ethanol to react as a nucleophile. As a result, ethyl acetate—the ester of ethanol and acetic acid—is the most abundant ester in wines. Other combinations of organic alcohols (such as phenol-containing compounds) and organic acids lead to a variety of different esters in wines, contributing to their different flavours, smells and tastes. Of course, when compared to sulfuric acid conditions, the acid conditions in a wine are mild, so yield is low (often in tenths or hundredths of a percentage point by volume) and take years for ester to accumulate.

Variations

Tetrabutylammonium tribromide (TBATB) can serve as an effective but unconventional catalyst for this reaction. It is believed that hydrobromic acid released by TBATB protonates the alcohol rather than the carboxylic acid, making the carboxylate the actual nucleophile. This would be a reversal of the standard esterification mechanism. An example of this method is the acylation 3-phenylpropanol using glacial acetic acid and TBATB. The reaction generates the ester in 15 minutes in a 95% yield without the need to remove water.

Glycerol

Glycerol

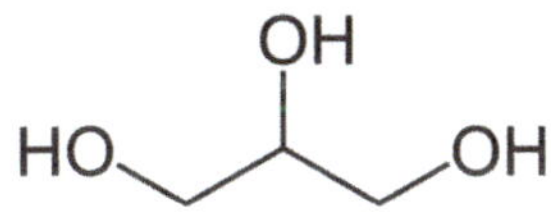

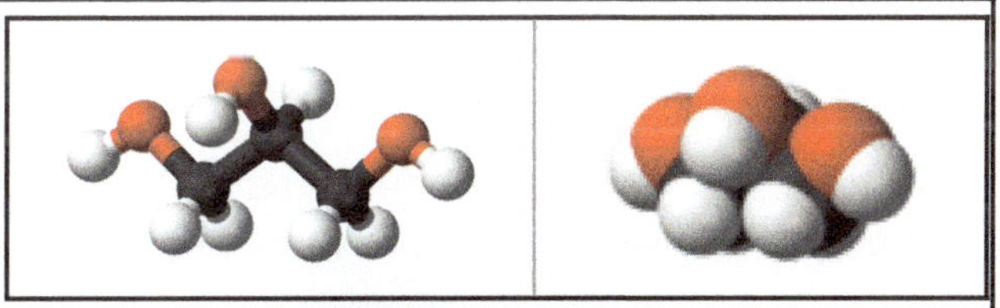

Glycerol is a simple polyol compound. It is a colorless, odorless, viscous liquid that is sweet-tasting and non-toxic. The glycerol backbone is found in all lipids known as triglycerides. It is widely used in the food industry as a sweetener and humectant and in pharmaceutical formulations. Glycerol has three hydroxyl groups that are responsible for its solubility in water and its hygroscopic nature.

Structure

Although achiral, glycerol is prochiral with respect to reactions of one of the two primary alcohols. Thus, in substituted derivatives, the stereospecific numbering labels each carbon as either sn-1, sn-2, or sn-3.

Production

Glycerol is generally obtained from plant and animal sources where it occurs as triglycerides. Triglycerides are esters of glycerol with long-chain carboxylic acids. The hydrolysis, saponification or transesterification of these triglycerides produces glycerol as well as the fatty acid derivative:.

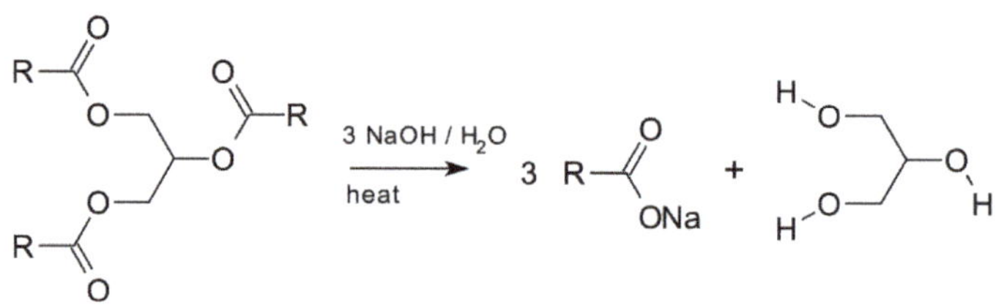

Triglycerides (1) are treated with an alcohol such as ethanol (2) with catalytic base to give ethyl esters of fatty acids (3) and glycerol (4):

Typical plant sources include soybeans or palm. Animal-derived tallow is another source. Approximately 950,000 tons per year are produced in the United States and Europe; 350,000 tons of glycerol were produced per year in the United States alone from 2000 to 2004. Production will increase as the EU directive 2003/30/EC are implemented, which required the replacement of 5.75% of petroleum fuels with biofuel across all member states by 2010, as glycerol is a byproduct in the production of biodiesel. It was projected in 2006 that by the year 2020, production would be six times more than demand.

Glycerol from triglycerides is produced on a large scale, but the crude product is of variable quality, with a low selling price of as low as 1–8 U.S. cents per pound in 2011. It can be purified, but the process is expensive. Some glycerol is burned for energy, but the heat value is low.

Crude glycerol from the hydrolysis of triglycerides can be purified by treatment with activated carbon to remove organic impurities, alkali to remove unreacted glycerol esters, and ion exchange to remove salts. High purity glycerol (> 99.5%) is obtained by multi-step distillation; vacuum is helpful due to the high boiling point of glycerol (290 °C).

Synthetic Glycerol

Although usually not cost-effective, glycerol can be produced by various routes from propylene. The epichlorohydrin process is the most important; it involves the chlorination of propylene to give allyl chloride, which is oxidized with hypochlorite to dichlorohydrins, which reacts with a strong base to give epichlorohydrin. This epichlorohydrin is then hydrolyzed to give glycerol. Chlorine-free processes from propylene include the synthesis of glycerol from acrolein and propylene oxide.

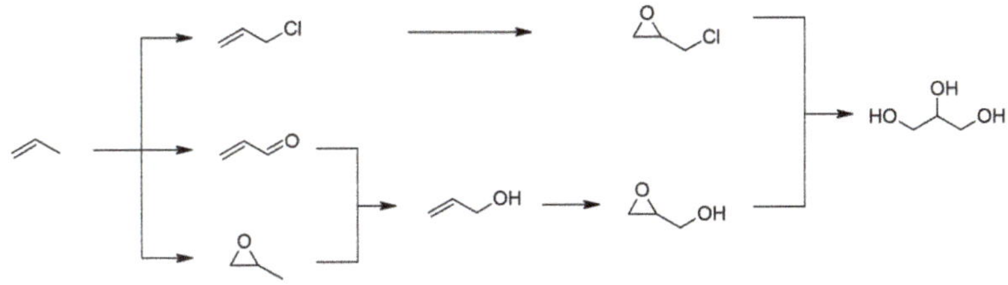

Because of the large scale production of biodiesel from fats, where glycerol is a waste product, the market for glycerol is depressed. Thus, synthetic processes are not economical. Owing to oversupply, efforts are being made to convert glycerol to synthetic precursors, such as acrolein and epichlorohydrin.

Applications

Food Industry

In food and beverages, glycerol serves as a humectant, solvent, and sweetener, and may help preserve foods. It is also used as filler in commercially prepared low-fat foods (e.g., cookies), and as a thickening agent in liqueurs. Glycerol and water are used to preserve certain types of plant leaves. As a sugar substitute, it has approximately 27 kilocalories per teaspoon (sugar has 20) and is 60% as sweet as sucrose. It does not feed the bacteria that form plaques and cause dental cavities. As a food additive, glycerol is labeled as E number E422. It is added to icing (frosting) to prevent it from setting too hard.

As used in foods, glycerol is categorized by the Academy of Nutrition and Dietetics as a carbohydrate. The U.S. Food and Drug Administration (FDA) carbohydrate designation includes all caloric macronutrients excluding protein and fat. Glycerol has a caloric density similar to table sugar, but a lower glycemic index and different metabolic pathway within the body, so some dietary advocates accept glycerol as a sweetener compatible with low carbohydrate diets.

It is also recommended as an additive when using polyol sweeteners such as erythritol and xylitol which have a cooling effect, due to its heating effect in the mouth, if the cooling effect is not wanted.

Pharmaceutical and Personal Care Applications

Glycerol is used in medical and pharmaceutical and personal care preparations, mainly as a means of improving smoothness, providing lubrication and as a humectant. It is found in allergen immunotherapies, cough syrups, elixirs and expectorants, toothpaste, mouthwashes, skin care products, shaving cream, hair care products, soaps and water-based personal lubricants. In solid dosage forms like tablets, glycerol is used as a tablet holding agent. For human consumption, glycerol is classified by the U.S. FDA among the sugar alcohols as a caloric macronutrient.

Glycerol suppositories used as laxatives

Glycerol is a component of glycerin soap. Essential oils are added for fragrance. This kind of soap is used by people with sensitive, easily irritated skin because it prevents skin dryness with its moisturizing properties. It draws moisture up through skin layers and slows or prevents excessive drying and evaporation.

Glycerol can be used as a laxative when introduced into the rectum in suppository or small-volume (2–10 ml) (enema) form; it irritates the anal mucosa and induces a hyperosmotic effect.

Taken orally (often mixed with fruit juice to reduce its sweet taste), glycerol can cause a rapid, temporary decrease in the internal pressure of the eye. This can be useful for the initial emergency treatment of severely elevated eye pressure.

Botanical Extracts

When utilized in "tincture" method extractions, specifically as a 10% solution, glycerol prevents tannins from precipitating in ethanol extracts of plants (tinctures). It is also used as an "alcohol-free" alternative to ethanol as a solvent in preparing herbal extractions. It is less extractive when utilized in a standard tincture methodology. Alcohol-based tinctures can also have the alcohol removed and replaced with glycerol for its preserving properties. Such products are not "alcohol-free" in a scientific sense, as glycerol contains three hydroxyl groups. Fluid extract manufacturers often extract herbs in hot water before adding glycerol to make glycerites.

When used as a primary "true" alcohol-free botanical extraction solvent in non-tincture based methodologies, glycerol has been shown to possess a high degree of extractive versatility for botanicals including removal of numerous constituents and complex compounds, with an extractive power that can rival that of alcohol and water/alcohol solutions. That glycerol possesses such high extractive power assumes it is utilized with dynamic methodologies as opposed to standard passive "tincturing" methodologies that are better suited to alcohol. Glycerol possesses the intrinsic property of not denaturing or rendering a botanical's constituents inert (as alcohols – i.e. ethyl (grain) alcohol, methyl (wood) alcohol, etc., do). Glycerol is a stable preserving agent for botanical extracts that, when utilized in proper concentrations in an extraction solvent base, does not allow inverting or reduction-oxidation of a finished extract's constituents, even over several years. Both glycerol and ethanol are viable preserving agents. Glycerol is bacteriostatic in its action, and ethanol is bactericidal in its action.

Electronic Cigarette Liquid

Vegetable glycerin is a common component of e-liquid, a solution used with electronic vaporizers (electronic cigarettes), that is heated with an atomizer to produce a vapor in order to deliver flavors and optionally nicotine.

Antifreeze

Like ethylene glycol and propylene glycol, glycerol is a non-ionic kosmotrope that forms strong hydrogen bonds with water molecules, competing with water-water hydrogen bonds. This disrupts the crystal lattice formation of ice unless the temperature is significantly lowered. The minimum freezing point temperature is at about −36 °F / −37.8 °C corresponding to 70% glycerol in water.

Glycerol was historically used as an anti-freeze for automotive applications before being replaced by ethylene glycol, which has a lower freezing point. While the minimum freezing point of a glycerol-water mixture is higher than an ethylene glycol-water mixture, glycerol is not toxic and is being re-examined for use in automotive applications.

In the laboratory, glycerol is a common component of solvents for enzymatic reagents stored at temperatures below 0 °C due to the depression of the freezing temperature. It is also used as a cryoprotectant where the glycerol is dissolved in water to reduce damage by ice crystals to laboratory organisms that are stored in frozen solutions, such as bacteria, nematodes, and mammalian embryos.

Internal Combustion Fuel

Used to power diesel generators supplying electricity for the FIA Formula E series of electric race cars.

Chemical Intermediate

Glycerol is used to produce nitroglycerin, which is an essential ingredient of various explosives such as dynamite, gelignite, and propellants like cordite. Reliance on soap-making to supply co-product glycerol made it difficult to increase production to meet wartime demand. Hence, synthetic glycerol processes were national defense priorities in the days leading up to World War II. Nitroglycerin, also known as glyceryl trinitrate (GTN) is commonly used to relieve *angina pectoris*, taken in the form of sub-lingual tablets, or as an aerosol spray.

Allyl iodide, a chemical building block for polymers, preservatives, organometallic catalysts, and pharmaceuticals, can be synthesized by using elemental phosphorus and iodine on glycerol.

A great deal of research is being conducted to try to make value-added products from crude glycerol (typically containing 20% water and residual esterification catalyst) obtained from biodiesel production. The use of crude glycerol as an additive to biomass for a renewable energy source when burned or gasified is also being explored.

- Hydrogen gas production unit
- Glycerine acetate (as a potential fuel additive)
- Conversion to propylene glycol
- Conversion to acrolein
- Conversion to ethanol
- Conversion to epichlorohydrin, a raw material for epoxy resins

Film Industry

Glycerol is used by the film industry when filming scenes involving water in order to stop areas from drying out too quickly.

Metabolism

Glycerol is a precursor for synthesis of triacylglycerols and of phospholipids in the liver and adipose tissue. When the body uses stored fat as a source of energy, glycerol and fatty acids are released into the bloodstream. Circulating glycerol does not glycate proteins as do glucose or fructose, and does not lead to the formation of advanced glycation endproducts (AGEs). In some organisms, the glycerol component can enter the glycolysis pathway directly and, thus, provide energy for cellular metabolism (or, potentially, be converted to glucose through gluconeogenesis).

Before glycerol can enter the pathway of glycolysis or gluconeogenesis (depending on physiological conditions), it must be converted to their intermediate glyceraldehyde 3-phosphate in the following steps:

Glycerol	Glycerol kinase	Glycerol-3-phosphate	Glycerol-3-phosphate dehydrogenase	Dihydroxyacetone phosphate	Triosephosphate isomerase	Glyceraldehyde 3-phosphate

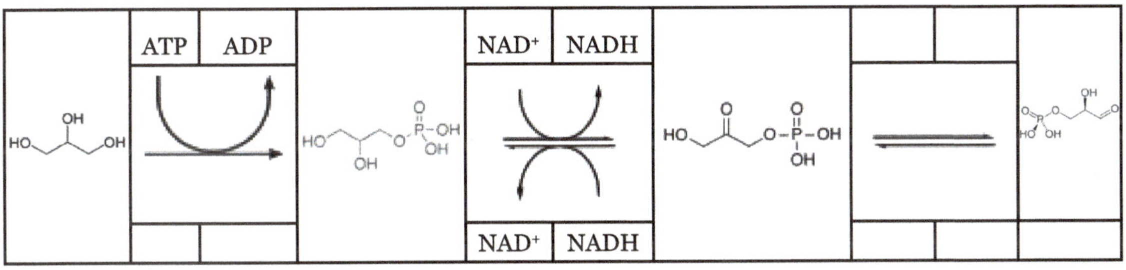

The enzyme glycerol kinase is present mainly in the liver and kidneys, but also in other body tissues, including muscle and brain. In adipose tissue, glycerol 3-phosphate is obtained from dihydroxyacetone phosphate (DHAP) with the enzyme glycerol-3-phosphate dehydrogenase.

Glycerol has very low toxicity when ingested; its LD_{50} oral dose for rats is 12600 mg/kg and 8700 mg/kg for mice.

Historical Cases of Contamination with Diethylene

On 4 May 2007, the US Food and Drug Administration advised all US makers of medicines to test all batches of glycerol for the toxic diethylene glycol. This followed an occurrence of hundreds of fatal poisonings in Panama resulting from a falsified import customs declaration by Panamanian import/export firm Aduanas Javier de Gracia Express, S. A. The cheaper diethylene glycol was relabeled as the more expensive glycerol.

Etymology

The origin of the gly- and glu- prefixes for glycols and sugars is from Greek *glukus* which means sweet.

References

- Homer-Dixon, Thomas (2007). The Upside of Down; Catastrophe, Creativity and the Renewal of Civilisation. Island Press. ISBN 978-1-59726-630-7.
- Furniss, Brian; Hannaford, Antony; Smith, Peter; Tatchell, Austin (1996). Vogel's Textbook of Practical Organic Chemistry 5th Ed. London: Longman Science & Technical. pp. 695–697 & 699–704. ISBN 9780582462366.
- Tingle, J. Bishop (1906). "Reports: Esterification". American Chemical Journal. 35: 368–369. Retrieved 19 January 2016.
- Wald, Matthew L. (2013-05-24). "USEC to Shut Uranium-Enrichment Plant in Kentucky". The New York Times. ISSN 0362-4331. Retrieved 2015-12-04.
- Ayres, Chris (27 February 2010) Clean green finish that sends a loved one down the drain Times Online. Retrieved 2013-02-20.
- "aluminium_water_hydrogen.pdf (application/pdf Object)" (PDF). www1.eere.energy.gov. 2008. Retrieved January 15, 2013.

- Andrew Hough (18 October 2012). "British engineers produce amazing 'petrol from air' technology". The Telegraph. Retrieved 21 October 2012.
- "Olives: Safe Methods for Home Pickling (application/pdf Object)" (PDF). ucanr.org. 2010. Retrieved January 22, 2012.
- Green, Margaret (January 1952). "A RAPID METHOD FOR CLEARING AND STAINING SPECIMENS FOR THE DEMONSTRATION OF BONE" (PDF). The Ohio Journal of Science. 52 (1): 31–33. Retrieved 20 November 2012.

Permissions

All chapters in this book are published with permission under the Creative Commons Attribution Share Alike License or equivalent. Every chapter published in this book has been scrutinized by our experts. Their significance has been extensively debated. The topics covered herein carry significant information for a comprehensive understanding. They may even be implemented as practical applications or may be referred to as a beginning point for further studies.

We would like to thank the editorial team for lending their expertise to make the book truly unique. They have played a crucial role in the development of this book. Without their invaluable contributions this book wouldn't have been possible. They have made vital efforts to compile up to date information on the varied aspects of this subject to make this book a valuable addition to the collection of many professionals and students.

This book was conceptualized with the vision of imparting up-to-date and integrated information in this field. To ensure the same, a matchless editorial board was set up. Every individual on the board went through rigorous rounds of assessment to prove their worth. After which they invested a large part of their time researching and compiling the most relevant data for our readers.

The editorial board has been involved in producing this book since its inception. They have spent rigorous hours researching and exploring the diverse topics which have resulted in the successful publishing of this book. They have passed on their knowledge of decades through this book. To expedite this challenging task, the publisher supported the team at every step. A small team of assistant editors was also appointed to further simplify the editing procedure and attain best results for the readers.

Apart from the editorial board, the designing team has also invested a significant amount of their time in understanding the subject and creating the most relevant covers. They scrutinized every image to scout for the most suitable representation of the subject and create an appropriate cover for the book.

The publishing team has been an ardent support to the editorial, designing and production team. Their endless efforts to recruit the best for this project, has resulted in the accomplishment of this book. They are a veteran in the field of academics and their pool of knowledge is as vast as their experience in printing. Their expertise and guidance has proved useful at every step. Their uncompromising quality standards have made this book an exceptional effort. Their encouragement from time to time has been an inspiration for everyone.

The publisher and the editorial board hope that this book will prove to be a valuable piece of knowledge for students, practitioners and scholars across the globe.

Index

U

V